Lecture Notes
in Control and Information Sciences   313

Editors: M. Thoma · M. Morari

Zhengguo Li · Yengchai Soh · Changyun Wen

Switched and Impulsive Systems

Analysis, Design, and Applications

With 129 Figures

 Springer

Authors

Prof. Zhengguo Li
Prof. Yengchai Soh
Prof. Changyun Wen

Nanyang Technological University
School of Electrical and Electronic Engineering
Block S2, Nanyang Avenue 50
639798 Singapore
Singapore

ISSN 0170-8643

ISBN 3-540-23952-9 **Springer Berlin Heidelberg New York**

Library of Congress Control Number: 2005920064

Springer is a part of Springer Science+Business Media

springeronline.com

© Springer-Verlag Berlin Heidelberg 2005
Printed in The Netherlands

Typesetting: Data conversion by the authors.
Final processing by PTP-Berlin Protago-TeX-Production GmbH, Germany
Cover-Design: design & production GmbH, Heidelberg
Printed on acid-free paper   89/3141Yu - 5 4 3 2 1 0

This work is dedicated to

Our Parents, and

Fang Xu and Anqi Li
— Zhengguo Li

Sokfong Lim, Hwaloong Soh, Hwahuey Soh, Hwajie Soh
and Hwaxiong Soh
— Yengchai Soh

Xiu Zhou, Wen Wen, Wendy Wen, Qingyun Wen
and Qinghao Wen
— Changyun Wen

Preface

Feedback and adaptation are two of the most important mechanisms whereby systems in nature sense adapt to the environment and to achieve certain objectives. These two mechanisms/concepts have also been widely used in the engineering system and computer science so that these systems can provide desired performance when subject to disturbance and/or changing circumstances. In this book, we shall discuss another very important concept of switched and impulsive control, which we believe will also be widely used in the analysis and control of complex systems.

A switched and impulsive system is a collection of finite continuous variable systems (CVSs) along with a discrete event system (DES) governing the impulsive "switching" among them. There are indeed many switched and impulsive systems that occur naturally or by design, in the fields of control, communication, computer and signal processes. Some examples are switched/server systems, switched flow network, Chua's circuit, computer disk drives, intelligent vehicle/highway systems, mobile robots of Hilare type, an Internet router, a scalable video coding system, a chaotic based secure communication system, and so on. Indeed, switched and impulsive systems have numerous applications in fields like mechanical systems, automotive industry, aircraft, air traffic control, network control, chaotic based secure communication, quality of service in the Internet, and video coding. This book aims to provide readers with a comprehensive coverage of switched and impulsive systems.

A novel method, called the *cycle analysis method*, and a new notion of redundancy of each cycle, are introduced for the stability analysis of switched and impulsive systems. We shall provide some conditions where the redundancy of each cycle in the switched and impulsive system can be removed. As a result, these conditions only require Lyapunov functions to be non-increasing along each type of cycle and to be bounded by a continuous function along each CVS. Since the numbers of cycles and CVSs are finite, these conditions can be checked easily. We shall also present the methods of Lyapunov functions, matrix norms, matrix measures, linear approximation methods and generalized matrix measures to measure the redundancy of each cycle and to study the stability of switched and impulsive systems. More importantly, with our results, a less conservative stability result for the switched and impulsive systems can be derived.

As an illustrative application, we derive less conservative conditions for the stabilization and synchronization of a class of chaotic systems by switched and/or impulsive control. With the scheme, the time necessary to synchronize two chaotic systems can be calculated and minimized while the bound of the impulsive interval after these two systems are synchronized is maximized. These results are extremely useful for the design of chaos based secure communication systems. With a larger impulsive interval, the transmission efficiency of chaotic secure communication systems is improved because less bandwidth is needed to transmit synchronization impulses. We shall also introduce a simple switched sampling mechanism for sampling chaotic systems so as to produce a flatter power spectrum. This will produce a more random set of key sequences, and hence it improves the security of the chaos-based cryptosystems.

Three practical applications of switched systems are also examined. A novel scheduling method, called the *simple cyclic control policy*, and its improved version, *feedback cyclic control policy*, are proposed to study the scheduling problem of a class of client/server systems that are composed of several clients and one server. Our method outperforms a popular method, called the *deficit round robin (DRR) method*, in the sense that neither admission control nor resource reservation is required by our method. Our methods can be generalized to study the scheduling problem of any switched flow network with a single server.

The *feedback cyclic control policy* is also extended to the case where the arrival rates are not known. The extended policy is then used to provide a relative differentiated quality of service in the Internet. The scheduling method is much simpler than the existing ones. It requires minimal changes to the current Internet infrastructure and therefore can be easily implemented to improve the performance of the current Internet. A novel source adaptation scheme and an adaptive media playout scheme are proposed for the Internet with the relative differentiated quality of service using the switched control. A very low cost solution is thus provided to transmit video over the Internet. The solution will be preferred by many cost conscious customers, especially students.

The methodology on switched system can also be used to study the problem of scalable video coding (SVC). The states of an SVC system are the given bit rate, resolution and frame rate, the control inputs are the motion information and residual data to be coded. A switched SVC scheme is proposed such that a customer oriented SVC scheme can be designed. It is focused on the trade-off between the motion information and the residual information. This is most crucial for an SVC scheme and is achieved by rate distortion optimization (RDO) with the utilization of a Lagrangian multiplier, which is adaptive to the customer composition in our scheme. A novel coding scheme for the SVC, i.e. cross layer motion estimation/motion compensation (ME/MC) scheme, is also proposed with the introduction of one new criterion to the SNR scalability and the spatial scalability, respectively, and a simple motion information truncation method is presented. Meanwhile, the full motion information scalability is provided by defining a switching law for the motion information and residual data. The coding efficiency is improved significantly with our scheme.

The book can be used as a reference or a text for a course at the graduate level. It is also suitable for self–study and for industry–oriented courses. The knowledge background for this monograph would be some undergraduate and graduate courses on linear system theory, graph theory, nonlinear system theory, basic network knowledge, basic cryptography knowledge, and basic video coding technology.

There are totally ten chapters in this monograph. Chapter 1 contains essential concepts and modelling of switched and impulsive systems. Numerous examples are provided in this chapter to illustrate the concept and applications of switched and impulsive systems. The next four chapters, i.e. Chapters 2-5, provide new theoretical developments for the analysis and design of switched and impulsive systems, followed by five chapters on the practical applications of switched and impulsive systems. Chapters 1 and 2 present the fundamentals for the whole book. It is advisable that you read them before you go through other chapters according to your interest. If you are interested in the analysis and design of switched and impulsive systems, you can proceed to read Chapters 3-5. For detail on the chaos based secure communication, you can directly go to Chapters 5 and 6. If you are more interested in the quality of service of the Internet, you can jump to Chapters 7, 8 and 10. If you are looking for applications in the video on demand, you just need to read Chapters 8-10.

The authors are grateful to Dr. Wenxiang Xie of Seagate Technology International, Singapore, Dr. Kun Li and Mr. Cheng Chen of School of Electrical and Electronic Engineering, Nanyang Technological University, for their helpful discussion. The first author would like to thank Prof. Xinhe Xu of Northeastern University, China for his guidance when he was a postgraduate student in China. All three authors would like to thank the editorial and production staff, especially Dr. Thomas Ditzinger, Ms. Heather King, Dr. Dieter Merkle and Dr. Riedesel, of Springer–Verlag, for their kind assistance in preparing this monograph.

Last but not least, Zhengguo Li would like to thank his wife Fang and daughter Anqi. Yengchai Soh would like to thank his wife Sokfong, and his sons Hwaloong, Hwahuey, Hwajie and Hwaxiong. Changyun Wen would like to thank his wife Xiu, his daughters Wen, Wendy and Qingyun and his son Qinghao.

Singapore,
October 2004

Zhengguo Li,
Yengchai Soh,
Changyun Wen

Contents

Symbols and Acronyms

R	the field of real numbers
R^+	$[0, \infty)$
$m!$	$m \times (m-1) \times \cdots \times 2 \times 1$
t_j	the jth switching instance of the system
t^-	the time just before τ
t^+	the time just after τ
$t_{s,i}^k$	the kth starting time of CVDS i
$t_{f,i}^k$	the kth finish time of CVDS i
$\Delta_{1,i}$	$\inf_{0 \le k < \infty} \{t_{f,i}^k - t_{s,i}^k\}$; if mode i is stable
$\Delta_{2,i}$	$\sup_{0 \le k < \infty} \{t_{f,i}^k - t_{s,i}^k\}$; if mode i is unstable
$[a, b]$	the closed interval
$C[R^+, R^+]$	the continuous function from R^+ to R^+
R^r	r dimensional real vector
X^T	the transpose vector of vector X
$\|X\|^2$	$X^T X$
$R^{r \times r}$	$r \times r$ dimensional real matrix
I	the identity matrix
A^T	the transpose matrix of matrix A
A^+	a right inverse matrix of A, which satisfies $AA^+ = I$
A^-	a left inverse matrix of A, which satisfies $A^- A = I$
$R^{n \times m}$	$n \times m$ matrices with real elements
$\prod_{l=1}^{n} A(l)$	$A(n)A(n-1)...A(1)$
$\sum_{l=1}^{n} A(l)$	$A(n) + A(n-1) + \cdots + A(1)$
A^{-1}	the inverse matrix of A
$\lambda(A)$	the eigenvalues of matrix A
$\lambda_{max}(A)$	the eigenvalue of matrix A whose real part is the maximum
$\lambda_{min}(A)$	the eigenvalue of matrix A whose real part is the minimum
$\rho(A)$	the spectral radius of matrix A

$\det(A)$ the determinant of matrix A

$\|A\|$ $\sqrt{\lambda_{max}(A^T A)}$

$a \bigoplus B$ $\max\{a, b\}$; $\forall a, b \in R$

$\|u(t_0, t)\|$ ess $\sup\{\|u(s)\|, s \in [t_0, t]\}$

$\|h(t)\|_\lambda$ $\sup_{t \in [0,T]} e^{-\lambda t}\|h(t)\|$; $h : [0, T] \to R$

$\prod_{k=l_1}^{l_2} A(k)$ $\begin{cases} I & l_1 > l_2 \\ A(l_2)A(l_2 - 1) \cdots A(l_1) & l_2 \geq l_1 \end{cases}$

$\mu(A)$ $\lambda_{max}[(A^T + A)/2]$

Class K $\{a \in C[R^+, R^+]$

    $a(t)$ is strictly increasing in t and $a(0) = 0\}$

Class CK $\{a \in C[R^+ \times R^+, R^+]$; $a(t, s) \in K$ for each t$\}$

$\forall$ for all

$\exists$ there exists

$\in$ belongs to

$\subset$ subset

$\cup$ union

$\cap$ intersect

$\sum$ sum

δ_A the maximum singular value of matrix A

$\underline{\delta}_A$ the minimum singular value of matrix A

CVS   continuous variable systems

DES   discrete event systems

ISS   input to state stability

QoS   quality of service

DRR   deficit round robin

FCCP   feedback cyclic control policy

RTT   round trip time

TCP   transmission control protocop

IP   Interenet protocol

SVC   scalable video coding

DCT   discrete cosine transform

DWT   discrete wavelet transform

UEP   unequal error protection

SNR   signal to noise ratio

PSNR   peak signal to noise ratio

MTU   maximum transfer unit

FGS   fine granularity scalability

MCTF   motion compensated temporal filtering

RDO   rate distortion optimization

GOP   group of pictures

HVS   human visual system

CIF   common intermediate format, 352×288

QCIF   quarter common inermediate format, 176× 144

MB   macroblock, 16×16

MAD	mean absolution difference
PSNR	peak signal-to noise ratio
MSE	mean squared error
ME/MC	motion estimation/motion compensaiton
MV	motion vector field
MVF	motion vector field

1. Examples and Modelling of Switched and Impulsive Systems

When you begin to read this book, you may ask: "what is a switched and impulsive system?" This is a question that may be best answered through several illustrative examples. Thus, in this chapter, we shall present a number of examples of switched and impulsive systems, starting from very simple ones to more complex but important classes of practical switched and impulsive systems.

1.1 Simple Examples of Switched and Impulsive Systems

Example 1.1.1. **A Switched Server System with Arrival Rate Less than Service Rate**

Consider a system consisting of three buffers and one server. The server removes work from any selected buffer at unit rate, and the work arrives at buffer $i(i = 1, 2, 3)$ at a constant rate of p_i ($\sum_{i=1}^{3} p_i < 1$).

The above is a simple example of switched system. To illustrate the essence of switched and impulsive systems much more clearly, a switching law for the server can be further defined as follows:

Cyclic fixed time scheduling method: After the server removes the work from the selected buffer for a specified period, it switches to the subsequent selected buffer with a positive reset time. The switching of the server forms a cycle and it repeats itself within the cycle.

On one hand, if you study the system behavior at a low level, you will find that the whole system is composed of four continuous variable systems (CVSs) given as follows:

$$\text{CVS 1:} \quad \begin{cases} \dot{X}_1(t) = p_1 - 1 \\ \dot{X}_2(t) = p_2 \\ \dot{X}_3(t) = p_3 \end{cases},$$

$$\text{CVS 2:} \begin{cases} \dot{X}_1(t) = p_1 \\ \dot{X}_2(t) = p_2 - 1 \\ \dot{X}_3(t) = p_3 \end{cases},$$

$$\text{CVS 3:} \begin{cases} \dot{X}_1(t) = p_1 \\ \dot{X}_2(t) = p_2 \\ \dot{X}_3(t) = p_3 - 1 \end{cases},$$

$$\text{CVS 4:} \begin{cases} \dot{X}_1(t) = p_1 \\ \dot{X}_2(t) = p_2 \\ \dot{X}_3(t) = p_3 \end{cases},$$

where CVS $i(i = 1, 2, 3)$ corresponds to the process that the server removes work from buffer i, and CVS 4 corresponds to the process that the server switches from one buffer to another one, and where $X_i(t)(i = 1, 2, 3)$ is the work in buffer i at time t.

On the other hand, if you study the system's behavior at a higher level, you will find that the logical relationship among these four CVSs can be described by a discrete event system (DES), which is illustrated in Fig. 1.1.

$\square$

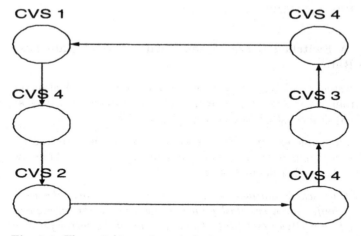

Fig. 1.1. The switchings of a switched server system

Example 1.1.2. **A DVD Burning System via a Universal Serial Bus (USB).**

Consider the problem of estimating the remaining time of a DVD burning system via a USB. Suppose that the size of a directory to be copied is $X_{1,0}$ and the initial speed of the USB is estimated as $X_{2,0}$. The estimation of the speed is updated every T seconds. The estimation of the remaining time, $X_3(t)$, is modelled as the following impulsive system:

$$\dot{X}_3(t) = -1 \, ; \, (k-1)T \le t < kT, \tag{1.1}$$

$$X_3(kT) = \frac{X_1(kT)}{X_2(kT)}, \tag{1.2}$$

where $X_1(kT)$ is the size of the remaining data to be copied, $X(0) = X_{1,0}$, $X_2(kT)$ is the estimation of the speed at time kT, and $X_2(0) = X_{2,0}$.

Let $X_4((k-1)T)$ be the total number of data that has been copied within the interval $[(k-1)T, kT)$. $X_1(kT)$ and $X_2(kT)$ are updated by

$$X_1(kT) = X_1((k-1)T) - X_4((k-1)T), \tag{1.3}$$

$$X_2(kT) = \mu X_2((k-1)T) + (1-\mu)\frac{X_4((k-1)T)}{T} \, ; \, 0 \le \mu < 1. \tag{1.4}$$

This is an example of impulsive system. Let $kT^+ = \lim_{\xi \to 0^+}(kT + \xi)$ and $kT^- = \lim_{\xi \to 0^+}(kT - \xi)$. The impulsive behavior of the above system is

$$X_3(kT^+) = X_3(kT^-) + \frac{X_1(kT)}{X_2(kT)} - \frac{X_1((k-1)T)}{X_2((k-1)T)} + T. \tag{1.5}$$

$\square$

Example 1.1.3. **A Router with Multiple Buffers in the Internet**

Consider an Internet where each router is connected to an adjacent router by two links: an outgoing link and an incoming link. The only outgoing link has a first-in-first-out (FIFO) buffer associated with each priority. This is illustrated in Fig. 1.2.

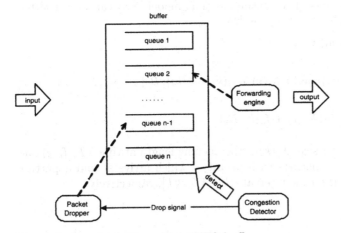

Fig. 1.2. A router with multiple FIFO buffers

Let n denote the total number of all priorities of traffic sources, and $X_i(t)$ the number of packets occupied in the ith priority buffer at time t. From the fluid-flow traffic model, we have the following n CVSs:

$$\begin{cases} \dot{X}_1(t) = f_1(t) \\ \quad\vdots \\ \dot{X}_i(t) = \begin{cases} f_i(t) - u_i(t); & X_i(t) > 0 \\ \max\{0, f_i(t) - u_i(t)\}; & X_i(t) = 0 \end{cases} ; \ i = 1, 2, \cdots, n \ , \\ \quad\vdots \\ \dot{X}_n(t) = f_n(t) \end{cases} \tag{1.6}$$

where $f_i(t)$ and $u_i(t)$ are the arrival rate and the service rate of the ith priority buffer at time t, respectively.

The switching law of these CVSs is given by

1. The router starts with buffer 1.

2. The router switches from buffer i to buffer $(i+1)$ at the kth times when buffer i is served for a specified time interval at the kth times for $i = 1, 2, \cdots, n - 1$.

3. The router switches back to buffer 1 at the $(k + 1)$ times when buffer n is served for a specified time interval at the kth times.

The above example is also a switched system, just like the switched server system. However, in this case, the arrival rate is unknown.

Suppose that an outgoing link has a transmission capacity of C (bits/s). Each outgoing link has a scheduler that monitors periodically the buffer occupancy level and computes the service rate of each priority buffer. Assume that the period is T seconds. The continuous time model (1.6) can be converted into the following discrete time model:

$$\begin{cases} X_1(k+1) = \min\{X_1(k) + f_1(k), B_s\} \\ \quad\vdots \\ X_i(k+1) = \min\{\max\{0, X_i(k) + f_i(k) - u_i(k)\}, B_s\} \ ; \ i = 1, 2, \cdots, n \ , \\ \quad\vdots \\ X_n(k+1) = \min\{X_n(k) + f_n(k), B_s\} \end{cases} \tag{1.7}$$

where B_s is the buffer size, $X_i(k)$ is the queuing length at time kT, $f_i(k)$ and $u_i(k)$ are the sizes of packets with priority i which arrive at and departure from the router in the time interval $((k - 1)T, kT]$, respectively.   □

1.2 Mathematical Modelling of Switched and Impulsive Systems

The process of mathematical modelling, from the physical phenomena to a model with mathematical descriptions, is essential in science and engineering. The dynamics of switched and impulsive systems can be understood by studying their mathematical descriptions. Mathematical relations typically involve the use of differential or difference equations to describe the behavior of each CVS, and finite automata or Petri nets to model the relationship among all CVSs. In this section, we shall provide a model to represent both the discrete and continuous properties of switched and impulsive systems.

As illustrated by earlier examples, a switched and impulsive system is composed of a finite number of CVSs:

$$\dot{X}(t) = f(X(t), m(t)) \tag{1.8}$$

where $X(t) \in R^r$ is the continuous valued component, $m(t) \in \bar{M} = \{1, \cdots, n\}$ is the discrete valued component which is left continuous with each value of $m(t)$, i, corresponding to a $f(\cdot, i)$ and $f : R^r \times \bar{M} \to R^r$ are continuously differentiable vector fields.

When the trajectory of system (1.8) meets the hypersurface

$$\hat{S}_{m(t_i^-), m(t_i^+)} = \{(X(t_i^-), t_i^-) | \phi(X(t_i^-), m(t_i^-), m(t_i^+), t_i^-) = 0\}, \tag{1.9}$$

impulsive "switchings" will happen as follows:

$$\begin{cases} X(t_i^+) = h(X(t_i^-), m(t_i^-), m(t_i^+)) \\ m(t_i^+) \in NM(m(t_i^-)) = \psi(m(t_i^-)) \end{cases} , \tag{1.10}$$

where $NM(m(t_i^-)) \subseteq \bar{M}$, $\phi : R^r \times \bar{M} \times \bar{M} \times R \to R, h : R^r \times \bar{M} \times \bar{M} \to R^r$ and $\psi : \bar{M} \to 2^{\bar{M}}$ where $2^{\bar{M}}$ is the set of all possible subsets of $\bar{M}$, ψ stands for a finite state machine [17].

Since our switched and impulsive system has a finite collection of n CVSs, i.e. $m(t) \in \{1, 2, \cdots, n\}$, we denote CVS i as mode i of the system. Then for CVS i, all the corresponding functions can be written as $f(X(t), i)$, $h(X(t), i, j)$, $\psi(i)$ etc.

The state of switched and impulsive systems is a 2–tuple of the form $S(t) = (m(t), X(t))$, where $m(t)$ is the discrete valued component and $X(t)$ is the continuous valued component.

Equations (1.9) and (1.10) describe the logical relationship along all CVSs, i.e. the discrete property of a switched and impulsive system. Special switched and impulsive systems are defined as below.

Definition 1.2.1. *A switched and impulsive system is said to be*

- *eventual quasi–periodic if there exist a pair of integers $N_h \le n$ and N_m, such that for any path $S(t_0)$, $S(t_1^+)$, $S(t_2^+)$, $\cdots$, $S(t_k^+)$, $\cdots$, we have*

$$m(t_{k+N_h}^+) = m(t_k^+) \; ; \; k \ge N_m. \tag{1.11}$$

- *eventual periodic if (1.11) holds and*

$$X(t_{k+N_h}^+) = X(t_k^+) \; ; \; k \ge N_m. \tag{1.12}$$

  $t_{k+N_h} - t_k$ is an eventual period of the switched system.
- *an impulsive system if the number of CVSs is 1, i.e. $n = 1$, and*

$$h(X(t_i^-), 1, 1) \ne X(t_i^-). \tag{1.13}$$

- *a switched system if*

$$h(X(t_i^-), m(t_i^-), m(t_i^+)) = X(t_i^-). \tag{1.14}$$

  holds for any pair of $m(t_i^-)$ and $m(t_i^+)$, and the number of CVSs is greater than 1, i.e. $n > 1$.
- *linear if both f and h are linear functions, i.e.*

$$f(X(t), m(t)) = A(m(t))X(t) + b(m(t)),$$
$$h(X(t_i^-), m(t_i^-), m(t_i^+)) = D(m(t_i^-), m(t_i^+))X(t_i^-) + e(m(t_i^-), m(t_i^+)).$$

- *nonlinear if either f or h is nonlinear.*   □

Definition 1.2.2. *The logical switchings of the system with $n > 1$ is said to be*

- *arbitrary if $NM(m(t_i^-)) = \bar{M}$ holds for all $m(t_i^-) \in \bar{M}$.*
- *governed by a finite state machine [17] if $m(t_i^+) \in \psi(m(t_i^-))$ also represents a finite state machine $(\bar{M}, E)$, where $E \subseteq \bar{M} \times \Sigma \times \bar{M}$ and Σ is the event set. It is assumed that there are no two edges going out from the same state associated with the same event, i.e. for the pair of edges in E*

$$[(i, e1, j), (i, e2, l) \in E \quad and \quad e1 = e2] \quad implies \quad j = l. \tag{1.15}$$

- *required to be defined if ψ cannot be given a priori.*   □

We now provide two practical examples which can be modelled as switched and impulsive systems.

Example 1.2.1. [94] **A Switched Server System with Arrival Rate Equal to Service Rate**

Consider a system consisting of three buffers and one server. The work arrives at each buffer at a constant rate of 1/3 the unit rate and the buffer removes work from any selected buffer at a unit rate. This is illustrated in Fig. 1.3.

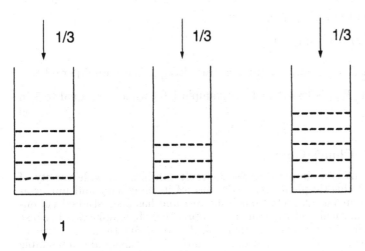

Fig. 1.3. A switched server system

Let $X_i(t)(i = 1, 2, 3)$ denote the amount of work in buffer i at time t. Then, we can obtain three CVSs

$$
\text{CVS 1:} \quad \begin{cases} \dot{X}_1(t) = -\frac{2}{3} \\ \dot{X}_2(t) = \frac{1}{3} \\ \dot{X}_3(t) = \frac{1}{3} \end{cases},
$$

$$
\text{CVS 2:} \quad \begin{cases} \dot{X}_1(t) = \frac{1}{3} \\ \dot{X}_2(t) = -\frac{2}{3} \\ \dot{X}_3(t) = \frac{1}{3} \end{cases},
$$

$$
\text{CVS 3:} \quad \begin{cases} \dot{X}_1(t) = \frac{1}{3} \\ \dot{X}_2(t) = \frac{1}{3} \\ \dot{X}_3(t) = -\frac{2}{3} \end{cases}.
$$

The switchings of these three CVSs are defined as

1. *The server starts with the first buffer.*

2. *If the server is in buffer j at time t, then the server remains there until the buffer is emptied.*

3. *When buffer $j (j = 1, 2)$ is empty, the server instantaneously switches to buffer $j + 1$. If buffer 3 is empty, the server then switches back instantaneously to buffer 1. The process is repeated.*

The switching condition sets and reset maps are

- $\hat{S}_{12} = \{X(t)|X_1(t) = 0\}$; $\hat{S}_{23} = \{X(t)|X_2(t) = 0\}$; $\hat{S}_{31} = \{X(t)|X_3(t) = 0\}$,
- $h(X, 1, 2) = h(X, 2, 3) = h(X, 3, 1) = X$,
- $\psi(1) = 2$; $\psi(2) = 3$; $\psi(3) = 1$.

The switched server system given in this example is eventual periodic.

The sum of all p_i is less than 1 in Example 1.1.1 while it is equal to 1 in this example. □

Example 1.2.2. **Chua's Circuit**

Chua's circuit is a simple electronic circuit exhibiting a wide variety of bifurcation and chaotic phenomena. Because of its simplicity and universality, Chua's circuit has attracted much interest and has been studied via numerical, mathematical and experimental approaches. It is universal because Chua's circuit has been proven mathematically to be chaotic in the sense of Shil'nikov's theorem [176]. It is simple because it contains only one simple nonlinear element and four linear elements [35, 37]. The dimensionless form of a Chua's circuit is

$$\begin{cases} \dot{x}(t) = \alpha(y(t) - x(t) - f(x(t))) \\ \dot{y}(t) = x(t) - y(t) + z(t) \\ \dot{z}(t) = -\beta y(t) - \gamma z(t) \end{cases}, \tag{1.16}$$

where $(x(t), y(t), z(t))$ is the state of Chua's circuit, $f(x(t))$ is the piecewise-linear characteristics of the Chua's diode and is given as

$$f(x(t)) = \vartheta_1 x(t) + \frac{1}{2}(\vartheta_2 - \vartheta_1)(|x(t) + 1| - |x(t) - 1|). \tag{1.17}$$

In (1.17), ϑ_1 and ϑ_2 are two constants and $\vartheta_2 < \vartheta_1 < 0$.

The Chua's circuit is a linear switched system that is composed of three CVSs:

$$\text{CVS 1: } A(1) = \begin{bmatrix} -(1+\vartheta_1)\alpha & \alpha & 0 \\ 1 & -1 & 1 \\ 0 & -\beta & -\gamma \end{bmatrix}, b(1) = \begin{bmatrix} (\vartheta_1 - \vartheta_2)\alpha \\ 0 \\ 0 \end{bmatrix},$$

$$\text{CVS 2: } A(2) = \begin{bmatrix} -(1+\vartheta_2)\alpha & \alpha & 0 \\ 1 & -1 & 1 \\ 0 & -\beta & -\gamma \end{bmatrix}, b(2) = \begin{bmatrix} 0 \\ 0 \\ 0 \end{bmatrix},$$

$$\text{CVS 3: } A(3) = \begin{bmatrix} -(1+\vartheta_1)\alpha & \alpha & 0 \\ 1 & -1 & 1 \\ 0 & -\beta & -\gamma \end{bmatrix}, b(3) = \begin{bmatrix} (\vartheta_2 - \vartheta_1)\alpha \\ 0 \\ 0 \end{bmatrix}.$$

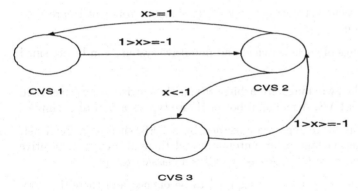

Fig. 1.4. The switchings of Chua's circuit

The switchings among them are illustrated in Fig. 1.4. The switching condition sets and reset maps are

- $\hat{S}_{12} = \{x(t)|x(t) \geq 1\}$; $\hat{S}_{21} = \{x(t)|1 > x(t) \geq -1\}$; $\hat{S}_{23} = \{x(t)|x(t) < -1\}$; $\hat{S}_{32} = \{x(t)|1 > x(t) \geq -1\}$,

- $h(X, 1, 2) = h(X, 2, 1) = h(X, 2, 3) = h(X, 3, 2) = X$,

- $\psi(1) = 2$; $\psi(2) = \{1, 3\}$; $\psi(3) = 2$.

$\square$

1.3 Control of Switched and Impulsive Systems

Essentially, a switched or an impulsive system is a collection of a finite number of CVSs along with maps that govern the impulsive "switching" among the CVSs [19]. The impulsive switchings occur whenever the states or the duration times of some CVSs satisfy certain conditions, respectively given by their memberships in the corresponding specified subsets of the state space. The impulsive switchings perform a reset to the active CVSs and change the continuous states of the CVSs. The switching relationships among all the CVSs can be represented by a discrete event system (DES). Hence, a switched or an impulsive system can be considered as a combination of a finite number of CVSs and a DES.

Generally, there are two cases in which a switched or an impulsive system can arise.

Case 1. The open-loop system is a switched or an impulsive system. Some typical examples include a caster mill [95], a switched server system

[94, 101], a switched flow network [122], a mobile robot of Hilare type [215] an so on.

The switchings of such switched and impulsive systems can be classified as follows:

Type 1. The switchings are arbitrary and there exists a positive dwell time for each CVS. A mobile robot of Hilare type is a typical example.

Type 2. The switchings are governed by a DES, such as a Petri net, a finite state machine or an automata, and there also exists a positive dwell time for each CVS. A caster mill is a classic example.

Type 3. The switchings are required to be defined and there does not exist any positive dwell time for such a switched or an impulsive system. In other words, the switchings can be arbitrarily fast. The switched server system is of this kind.

Case 2. The closed-loop system is a switched or an impulsive system. This class of switched and impulsive systems is generated when a group of continuous and/or impulsive controllers are designed to control a continuous process. At any time, an active continuous and/or impulsive controller, considered as a sub-controller, is selected based on certain performance indices to control the process. Such a group of sub-controllers, together with the corresponding switching law, forms a switched and/or impulsive controller. The switched and/or impulsive controller and the original continuous process form a switched and/or impulsive system.

Now impulsive control, switched control, and switched and impulsive control are formally defined respectively as follows:

Definition 1.3.1. *[206] Consider a plant P whose state variable is denoted by $X \in R^r$, a set of control instants $T = \{t_k\}, t_k \in R_+, t_k > t_{k-1}, k = 1, 2, \cdots$, and impulsive control laws $U(t_k, X) \in R^r, k = 1, 2, \cdots$. An impulsive control is defined as one in which at each t_k, $X(t)$ is changed impulsively, i.e. $X(t_k^+) = X(t_k^-) + U(t_k, X)$, such that the system is stable and certain specifications are achieved.* □

Definition 1.3.2. *[104] Consider a plant P whose state variable is denoted by $X \in R^r$. Suppose that we have a collection of state feedback controllers:*

$$\tilde{U}(t) = K_{m(t)}(X(t)) \; ; \; m(t) \in \{1, 2, \cdots, n\}, \tag{1.18}$$

where $K_i(i = 1, 2, \cdots, n) : R^r \to R^p$ are given continuous functions.

The controllers in (1.18) are called sub-controllers. A switched control is the law for switching one sub-controller to another one and is defined as

$$m(t) = I(X(t)), \tag{1.19}$$

where $I : R^r \to \{1, 2, \cdots, n\}$. □

Clearly, there is also a time sequence $\{t_k\}$ for a switched control where the sub-controller is switched from one to another at each t_k.

Definition 1.3.3. *Consider a plant P whose state variable is denoted by $X \in R^r$, a set of control instants $T = \{t_k\}, t_k \in R_+, t_k > t_{k-1}, k = 1, 2, \cdots$, impulsive control laws $U(t_k, X) \in R^r, k = 1, 2, \cdots$, and a collection of state feedback controllers (1.18). A switched and impulsive control is defined as one in which at each t_k, $X(t)$ is changed impulsively, i.e. $X(t_k^+) = X(t_k^-) + U(t_k, X)$, and the law for the switchings of sub-controllers is defined by (1.19).*

□

There are two main reasons to choose switched and/or impulsive control rather than continuous control. The first reason is that switched and/or impulsive control can be used to obtain better performance. For example, switched control have been used to achieve stability and to improve transient response of control systems in [138, 134, 84, 61, 140]. Another example is that the impulsive control can be used to improve the bandwidth utilization in the chaos-based secure communication, which have been updated to fourth generation where discontinuous or impulsive synchronization is employed [205, 109]. The continuous chaotic synchronization is adopted in the first three generations. Bandwidth of 30kHz is needed for transmitting the synchronization signals of a third-order chaotic transmitter in the first three generations of chaos-based cryptosystems, while less than bandwidth of 94 Hz is required in the fourth generation. Therefore, the efficiency of bandwidth usage is greatly improved by the impulsive controller. Indeed, switched and impulsive control has been employed in [109] for the synchronization of two identical chaotic systems. With the technique, the time necessary to synchronize two chaotic systems is minimized while the bound of the impulsive interval is maximized. Furthermore, a switched sampling of the chaotic signals can be employed to improve randomness of the generated key sequence. This greatly improve the security of chaos-based communications.

The second reason is that certain given objectives can be achieved only with the application of switched and/or impulsive control. An example is the feedback stabilization of underactuated mechanical systems. The systems are used for reducing weight, cost, or energy consumption, while still maintaining an adequate degree of dexterity without reducing the reachable configuration space. They also have the advantage of no or lesser damage when hitting an object, and are tolerant to the failure of actuators [217].

In the next section, some examples are used to illustrate the above ideas.

1.4 Practical Examples

It is not difficult to find practical examples to motivate the study of switched and impulsive systems. In this section, we provide many practical examples from the fields of manufacturing systems, chaotic secure communication, video coding and computer networks.

Example 1.4.1. **Synchronization of Two Identical Lorenz Systems via Impulsive Control**

The Lorenz system was first introduced as an approximate model of the unpredictable behavior of weather [48]. It is expanded from a set of non-linear partial differential equations using Fourier transformations and then truncated by retaining only three modes. The resulting equations, generally called Lorenz equations, consist of an autonomous nonlinear system of three ordinary differential equations. They are

$$\begin{cases} \dot{x}(t) = -\delta x(t) + \delta y(t) \\ \dot{y}(t) = \vartheta_3 x(t) - y(t) - x(t)z(t) \ , \\ \dot{z}(t) = x(t)y(t) - \vartheta_4 z(t) \end{cases} \tag{1.20}$$

where $(x(t), y(t), z(t))$ is the state of the Lorenz system, δ, ϑ_3 and ϑ_4 are positive numbers and represent the parameters of the Lorenz system.

There are two nonlinear equations in the Lorenz equations, which are functions of two variables, xz and xy, respectively, and there are three control parameters: δ, ϑ_3 and ϑ_4. The Lorenz system has been proposed for use in chaotic secure communication systems and chaotic spread spectrum communications [42, 54].

Let $X^T(t) = (x(t), y(t), z(t))$, then we can rewrite equation (1.20) as

$$\dot{X}(t) = AX(t) + \psi(X(t)), \tag{1.21}$$

where

$$A = \begin{bmatrix} -\delta & \delta & 0 \\ \vartheta_3 & -1 & 0 \\ 0 & 0 & -\vartheta_4 \end{bmatrix}, \tag{1.22}$$

$$\psi(X) = \begin{bmatrix} 0 \\ -xz \\ xy \end{bmatrix}. \tag{1.23}$$

In an impulsive synchronization configuration, the driving system is given by (1.20), whereas the driven system is

$$\dot{\tilde{X}}(t) = A\tilde{X}(t) + \psi(\tilde{X}(t)), \tag{1.24}$$

where $\tilde{X}(t) = (\tilde{x}(t), \tilde{y}(t), \tilde{z}(t))^T$ is the state variables of the driven system and A and ψ are as defined in (1.22) and (1.23).

At discrete instants $t_i (i = 1, 2, \cdots)$, the state variables of the driving system are transmitted to the driven system and the state variables of the driven system are then subject to jumps at these instants. In this sense, the driven system is modelled by the following impulsive equations:

$$\begin{cases} \dot{\tilde{X}}(t) = A\tilde{X}(t) + \psi(\tilde{X}(t)) \; ; \; t \neq t_k \\ \Delta\tilde{X}(t)|_{t=t_k} = -Be(t_k) \; ; \; k = 1, 2, \cdots \end{cases}, \qquad (1.25)$$

where B is a 3×3 symmetric matrix satisfying $\rho(I + B) < 1$, and $e^T(t) = (e_x(t), e_y(t), e_z(t)) = (x(t) - \tilde{x}(t), y(t) - \tilde{y}(t), z(t) - \tilde{z}(t))$ is the synchronization error.

Let

$$\Psi(X, \tilde{X}) = \psi(X) - \psi(\tilde{X}) = \begin{bmatrix} 0 \\ -(xz - \tilde{x}\tilde{z}) \\ xy - \tilde{x}\tilde{y} \end{bmatrix}. \qquad (1.26)$$

The error system of the impulsive synchronization is

$$\begin{cases} \dot{e}(t) = Ae(t) + \Psi(X, \tilde{X}) \; ; \; t \neq t_k \\ \Delta e(t)|_{t=t_k} = Be(t_k) \; ; \; k = 1, 2, \cdots \end{cases}. \qquad (1.27)$$

Clearly, we have an impulsive system. □

Example 1.4.2. Consider the population control problem, where the population of a country is the state vector $X(t)$ and the policy of the country as the control input $U(t)$. Suppose that the whole set of $X(t)$ is Ω and that Ω can be divided into n subsets $\Omega_i (i = 1, 2, \cdots, n)$, then

$$\tilde{U}(t) = K_i(X(t)) \quad \text{when} \quad X(t) \in \Omega_i. \qquad (1.28)$$

This formula is quite reasonable. When a country is short of manpower, the citizen will be encouraged to have more children. However, if the population of a country is too large, then the people should be restricted to have fewer children. □

Example 1.4.3. Consider the economic system of an underdeveloped country. Different economic laws should be applied in different situations given that the economic system of an underdeveloped country is not robust or strong enough to handle all possibilities.

For example, in the initial stage of development, the law can be "a cat, no matter white or black, is a good one if it can catch mice". After several years, it will switch to "a cat, no matter white or black, is a good one if it can catch mice for the whole country". Another five or ten years later, it will switch to "a cat, no matter white or black, is a good one if it can catch mice for the whole country legally". □

Example 1.4.4. Consider a scalable video coding system, with the given bit rate, resolution, and frame rate as the state vector $X(t)$, the motion information and residual data to be coded as the control input $U(t)$. Suppose that the whole set of $X(t)$ is Ω which can be divided into n subsets $\Omega_i (i = 1, 2, \cdots, n)$, then

$$\tilde{U}(t) = K_i(X(t)) \quad \text{when} \quad X(t) \in \Omega_i, \tag{1.29}$$

where the design of K_i will be discussed in Chapter 9.

Obviously, this is a switched scalable video coding system. Two major tasks should be completed to design the system.

Task 1 Within the Ω_i, the following two items should be determined for each frame:

- The motion information to be coded;

- The residual image to be coded.

Task 2 The switching point from one group of motion information and residual data to another group of motion information and residual data. □

There may not exists any dwell time for such switched and impulsive systems, that is, the switchings of the system can be arbitrary fast.

Example 1.4.5. **A Crossroad Scheduling System**

Consider a crossroad system illustrated in Fig. 1.5. A1, B1, C1 and D1 are the Red–Amber–Green signals and A2, B2, C2 and D2 stand for Turn-Left/No-Left-Turn signals.

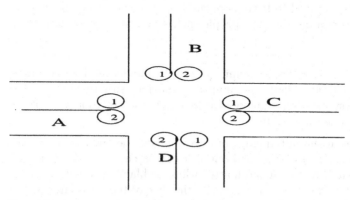

Fig. 1.5. A crossroad scheduling system

Signals A1, B1, C1, D1, A2, B2, C2 and D2 are used to control the traffic in a crossroad. Four different groups of signals are

1 A1–Green, B1–Red, C1–Green, D1–Red, B2–No-Left-Turn, D2–No-Left-Turn;
2 A1–Amber, B1–Red, C1–Amber, D1–Red, B2–Turn-Left, D2–Turn-Left;
3 A1–Red, B1–Green, C1–Red, D1–Green, A2–No-Left-Turn, C2–No-Left-Turn;
4 A1–Red, B1–Amber, C1–Red, D1–Amber, A2–Turn-Left, C2–Turn-Left.

The holding time of a special group of signals is set according to the traffic loads along the special direction at the crossroad. This is a quasi-periodic switched system. □

Example 1.4.6. A Switched Flow Network

Consider a network with $(n+1)$ nodes $\{NG_1, NG_2, \cdots, NG_n, NG_{n+1} = \infty\}$ where ∞ represents the exterior of the network. The edge departing from NG_i and arriving at NG_j is denoted by (NG_i, NG_j). Subsequently, any edge of the form (∞, NG_i) and (NG_i, ∞) $(i = 1, 2, \cdots, n)$ are regarded respectively as coming from and going to the outside of the system.

Suppose that the network can be divided into M layers [122]. There is a server, which removes the work from a selected buffer at the unit rate. Moreover, the work arrives to the first layer continuously at a constant rate of less than one unit rate. □

Both the switched server system and the switched flow network are switched and impulsive systems where the switchings of the system are required to be defined. At any time, there exists only one active CVS because there is only one server in the system.

Example 1.4.7. A Token Bus Protocol

A token bus protocol requires a control frame called an access token. This token gives a station the exclusive use of the bus. The token-holding station occupies the bus for a period of time to send data. It then passes the token to a designated station called the *successor station*. In bus topology, all stations listen to the channel and receive the access token, but the only station allowed to use the channel is the successor station. All other stations must wait for their turn to receive the token. □

Example 1.4.8. Context Adaptive Variable Length Coding (CAVLC) [8]

The CAVLC is an efficient entropy coding approach to code the transform coefficient levels of block-based motion compensated coding of video. There are seven VLC tables listed in Tables 1.1-1.7. VLC Tables 0 and 1 are primary and secondary tables, respectively.

The switching law among the tables is

The selection of table 0 or 1 for the first coefficient level is determined solely by the local variables representing the total number of non-zero coefficients and the number of trailing ones in the sequence of coefficient levels. The table for subsequent coefficient levels is determined solely by the previous coded coefficient level and an experimentally pre-determined table. □

Table 1.1. VLC Table 0

Code (bitstream bits)	Coefficient Level
1	1
01	-1
...	...
00000000000001	-7
000000000000001xxxs	±8 to ±15
0000000000000001	±16 and others

Table 1.2. VLC Table 1

Code No	Code (bitstream bits)	Coefficient Level
0-1	1s	±1
2-3	01s	±2
...	...	...
28-29	000000000000001s	±15
30 and above	0000000000000001xxxxxxxxxxxs	±16 and others

Table 1.3. VLC Table 2

Code No	Code (bitstream bits)	Coefficient Level
0-3	1xs	±1, ±2
4-7	01xs	±3, ±4
...	...	...
56-59	000000000000001xs	±29, ±30
60 and above	0000000000000001xxxxxxxxxxxs	±31 and others

Table 1.4. VLC Table 3

Code No	Code (bitstream bits)	Coefficient Level
0-7	1xxs	±1-±4
8-16	01xxs	±5-±8
...	...	...
112-119	000000000000001xxs	±57-±60
120 and above	0000000000000001xxxxxxxxxxxs	±61 and others

Table 1.5. VLC Table 4

Code No	Code (bitstream bits)	Coefficient Level
0-15	1xxxs	±1-±8
16-31	01xxxs	±9-±16
...	...	...
224-239	000000000000001xxxs	±113-±120
240 and above	0000000000000001xxxxxxxxxxxs	±121 and others

Table 1.6. VLC Table 5

Code No	Code (bitstream bits)	Coefficient Level
0-31	1xxxxs	±1-±16
32-63	01xxxxs	±17-±32
...	...	...
448-479	000000000000001xxxxs	±225-±240
480 and above	0000000000000001xxxxxxxxxxxs	±241 and others

Table 1.7. VLC Table 6

Code No	Code (bitstream bits)	Coefficient Level
0-63	1xxxxxs	± 1-± 32
64-127	01xxxxxs	± 33-± 64
...	...	...
896-959	000000000000001xxxxxs	± 449-± 480
960 and above	0000000000000001xxxxxxxxxxxxs	± 481 and others

Example 1.4.9. **A Switched Rate Control Scheme** [190]

Rate control strategy plays a critical role in video transmission because the communication channel often poses serious constraints on the available bit rate. Not surprisingly, the quality of the encoded video depends heavily on the rate control.

Assume that the discrete cosine transform (DCT) coefficients of the motion compensated difference frame are approximately uncorrelated and Laplacian distributed with variance σ^2. Suppose that X_1 and X_2 are the number of available bits and the corresponding quantization parameter for the current frame, respectively. The relationship between them is given by one of the following three equations:

$$\text{Model 1: } X_1 = e^{\alpha_1}\left(\frac{\sigma}{X_2}\right)^{\beta_1}, \tag{1.30}$$

$$\text{Model 2: } X_1 = e^{\alpha_2}\left(\frac{\sigma}{X_2}\right)^{\beta_2}, \tag{1.31}$$

$$\text{Model 3: } X_1 = e^{\alpha_3}\left(\frac{\sigma}{X_2}\right)^{\beta_3}. \tag{1.32}$$

With two predefined threshold values θ_{low} and θ_{high} satisfying $1 < \theta_{low} < \theta_{high} < 31$, the switching law between them is given by

If $1 \leq X_2 \leq \theta_{low}$, Model 1 is then chosen; Otherwise if $\theta_{low} \leq X_2 \leq \theta_{high}$, Model 2 is then selected; Otherwise, Model 3 is used.

It was shown in [190] that a switched rate control scheme provides a more accurate estimation of bit rate than existing models.   □

Other motivating examples of switched and impulsive systems are computer disk drives [53], stepper motors [23], constrained robotic systems [9], intelligent vehicle/highway systems [186] and a mobile robot of Hilare type [215].

1.5 Preview of Chapters

There are totally ten chapters in this monograph. Among them, four chapters (6-9) are dedicated to the practical applications of switched and impulsive systems and four chapters (2-5) to the new theoretical analysis of switched and impulsive systems. Brief description of chapters 2-10 are given below.

Chapter 2 formulates the state space of switched and impulsive systems and defines the stability of switched and impulsive systems with respect to an invariant set or an equilibrium. Less conservative stability conditions are derived in the sense that Lyapunov functions are only required to be non-increasing along a subsequence of the switchings and to be bounded by a continuous function along each CVS. The concept of cycle is also introduced for the stability analysis. With the conditions and the concept, it is possible to introduce a new notion, called the redundancy of each cycle, to study the stability of switched and impulsive systems.

Chapter 3 analyzes the stability of linear switched and impulsive systems. Three types of tools, namely Lyapunov functions, matrix norms and matrix measures are provided to measure the redundancy of each cycle. The *cycle analysis method* is applied to identify the non-increasing subsequence and to construct the continuous functions to bound Lyapunov functions along each CVS. A decomposition method is proposed to study the stability of linear switched and impulsive systems, which is totally composed of unstable CVSs. The results are also used to study the stabilization of bilinear systems.

Chapter 4 studies the stability of nonlinear switched and impulsive systems. Three types of methods, namely Lyapunov functions, linear approximation methods and generalized matrix measures, are presented to measure the redundancy of each cycle. Compared with the existing results, the results obtained in chapters 3 and 4 are less conservative and easier to be checked.

Chapters 5 and 6 provide less conservative conditions for the synchronization of chaotic systems via impulsive control, switched and impulsive control. The time necessary to synchronize two chaotic systems is minimized while the bound of the impulsive interval after two systems are synchronized is maximized. The transmission efficiency of the chaotic secure communication systems is improved significantly because less bandwidth is needed to transmit the synchronization impulses. Meanwhile, a concept of magnifying-glass and a novel switched sampling scheme are introduced to improve the security of the chaos cryptosystem. The proposed system can be applied to transmit text, speech, image files, and any digital binary data.

Chapter 7 provides a practical application of switched systems. A novel scheduling method, *simple cyclic control policy* [122] and its improved version, *feedback cyclic control policy*, are proposed to study the scheduling problem of a class of client/server systems. This chapter is emphasized on the case where the arrival rates are known in advance. Our method outperforms one of the most popular methods, the *deficit round robin (DRR)* method, in the sense that neither admission control nor resource reservation is required by our method. The policies can be generalized to study the scheduling problem of any switched flow network with a single server.

Chapter 8 presents another interesting application of switched systems. Specifically, the *feedback cyclic control policy* is firstly extended to the case that the arrival rates are not known in advance. It is then used to design a novel scheduling method, i.e. the *dual feedback cyclic control policy* to provide the relative differentiated quality of service in the current Internet. Each router, together with the dual feedback cyclic control policy, is a switched system. A source adaptation scheme and an adaptive media playout scheme are also presented for the relative differentiated quality of service using the switched control. A very low cost solution is provided to transmit video over the Internet by all of them. Many customers including students may choose this cheaper solution.

Chapter 9 proposes a switched scalable video (SVC) coding scheme by using the methodology on switched system. The states of an SVC system include the given bit rate, resolution and frame rate. The control inputs are the motion information and the residual data to be coded. This chapter focuses on the trade-off between the motion information and the residual data, which is most crucial for an SVC scheme. The tradeoff is achieved by rate distortion optimization (RDO) with the utilization of a Lagrangian multiplier. The Lagrangian multiplier is adaptive to the customer composition in our scheme. A novel coding scheme for the SVC, i.e. cross layer motion estimaiton/motion compensation (ME/MC) scheme, is also proposed with the introduction of one new criterion to the SNR scalability and the spatial scalability, respectively, and a simple motion information truncation method is presented. Meanwhile, the full motion information scalability is provided by defining a switching law for the motion information and residual data. The coding efficiency is improved significantly by using our scheme.

Chapter 10 highlights several more advanced applications and associated future research directions.

2. Analysis of Switched and Impulsive Systems

The analysis will focus on the case where the switchings are governed by a finite state machine. This is because the common Lyapunov function method may be the only way to study the stability of switched systems in the case where the switchings are arbitrary fast, and this topic has been well studied by many researchers. Narandra and Balakrishana[139], Shorten and Narandra [172, 173], Mori et al. [131, 130] and Ooba and Funahashi [144, 145] have presented sufficient conditions for the existence of such a common Lyapunov function and proposed an explicit construction of a quadratic common Lyapunov function for a finite commuting family of linear systems. Shim et al. [169] directly generalized the result and the proof technique of [139] to nonlinear switched systems.

2.1 The State Evolution of Switched and Impulsive Systems

To analyze switched and impulsive systems, the state evolution should be investigated. There are two types of changes, namely quantitative and qualitative changes. Both are constrained by $m(t)$ and $X(t)$. If no impulsive switching occurs, then a quantitative change only happens on $X(t)$. When an impulsive switching occurs, both $m(t)$ and $X(t)$ will change, and this is a qualitative change.

1. **The rules for quantitative changes are**

   - The rule from $X(t_{k-1}^+)$ to $X(t_k^-)$

$$X(t_k^-) = X(t_{k-1}^+) + \int_{t_{k-1}^+}^{t_k^-} f(X(t), m(t_{k-1}^+))dt. \tag{2.1}$$

   - The rule from $m(t_{k-1}^+)$ to $m(t_k^-)$

$$m(t_k^-) = m(t_{k-1}^+). \tag{2.2}$$

2. **The rules for qualitative changes are**

- The rule from $X(t_k^-)$ to $X(t_k^+)$

$$X(t_k^+) = h(X(t_k^-), m(t_k^-), m(t_k^+)).$$ (2.3)

- The rule from $m(t_k^-)$ to $m(t_k^+)$

$$m(t_k^+) \in \psi(m(t_k^-)).$$ (2.4)

where $k = 1, 2, \cdots, \infty$.

Example 2.1.1. Consider Example 1.2.1. Suppose that $m(t_0) = 1$ and $X(t_0) = [1, 0, 0]^T$. The rules for quantitative and qualitative changes are as follows:

1.1 The rules for quantitative changes with $t \in [0, 1.5)$ are

$$X(t) = [1 - \frac{2t}{3}, \frac{t}{3}, \frac{t}{3}]^T,$$
$$m(t) = 1.$$

1.2 The rules for qualitative changes at $t = 1.5$ are

$$X(1.5^+) = [0, 0.5, 0.5]^T,$$
$$m(1.5^+) = 2.$$

2.1 The rules for quantitative changes with $t \in [1.5, 2.25)$ are

$$X(t) = [\frac{t - 1.5}{3}, 0.5 - \frac{2t - 3}{3}, 0.5 + \frac{t - 1.5}{3}]^T,$$
$$m(t) = 2.$$

2.2 The rules for qualitative changes at $t = 2.25$ are

$$X(2.25^+) = [0.25, 0, 0.75]^T,$$
$$m(2.25^+) = 3.$$

3.1 The rules for quantitative changes with $t \in [2.25, 3.375)$ are

$$X(t) = [0.25 + \frac{t - 2.25}{3}, \frac{t - 2.25}{3}, 0.75 - \frac{2t - 4.5}{3}]^T,$$
$$m(t) = 3.$$

3.2 The rules for qualitative changes at $t = 3.375$ are

$$X(3.375^+) = [0625, 0.375, 0]^T,$$
$$m(3.375^+) = 1.$$

The state evolution of this example is illustrated in Fig. 2.1.   □

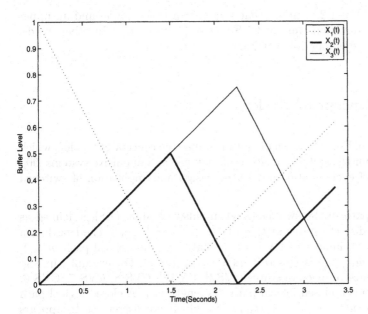

Fig. 2.1. The state evolution of a switched server system

It can be shown from Equations' (2.1)–(2.4) that when a switched and impulsive system evolves from an initial state $S(t_0)$, a switching sequence arises. Let it be denoted by

$$\Sigma = \{(m(t_0), t_0), (m(t_1^+), t_1) \cdots, m(t_k^+), t_k), \cdots, k \in N\}, \tag{2.5}$$

and let $\Sigma|_t$ denote the sequence of switching time when the model of the CVS is switched on, i.e.

$$\Sigma|_t = \{t_0, t_1^+, t_2^+, \cdots, t_k^+, \cdots\}. \tag{2.6}$$

There are infinite but countable "switchings" of a switched and impulsive system, and $t_k \to \infty$ as $k \to \infty$. Further, let $t_{s,i}^k$ be the k–th starting time of CVS i and $t_{f,i}^k$ be the k–th ending time of CVS i. For the case that the switchings of the system are governed by a finite state machine, notions of minimum and maximum duration times are introduced as follows [49, 63, 153, 10]:

$$0 < \Delta_{1,i} \triangleq \inf_{0 \leq k < \infty} \{t_{f,i}^k - t_{s,i}^k\}, \tag{2.7}$$

$$\Delta_{2,i} \triangleq \sup_{0 \leq k < \infty} \{t_{f,i}^k - t_{s,i}^k\} < \infty \; ; \; i = 1, 2, \cdots, n. \tag{2.8}$$

Equations (2.7) and 2.8) imply that there are a lower bound and an upper bound on the duration time of CVS i ($i \in \{1, 2, \cdots, n\}$). In other words, there are infinite number of switchings.

2.2 The Concept of Cycle

With the help of Fig. 2.2, we shall first introduce the concept of "cycle", which is very useful for the stability analysis of switched and impulsive systems [90]. The concept of cycle is also used in the design and application of switched systems.

A logical path in the switched and impulsive system (1.8) is defined as a sequence of discrete states $m(t_{j_1}), m(t_{j_1+1}), \cdots, m(t_{j_1+k})$. The length of a path is the total number of CVSs in the path. A finite logical path $m(t_{j_1})$, $m(t_{j_1+1}), \cdots, m(t_{j_1+k})$ is closed if $m(t_{j_1}) = m(t_{j_1+k})$. For example, in Fig. 2.2, the sequence of discrete states CVS 9, CVS 6, CVS 8, CVS 7, CVS 6, CVS 8, CVS 9 is a closed logical path. Its length is 7. A closed logical path $LC = m(t_{j_1}), m(t_{j_1+1}), \cdots, m(t_{j_1+k})$ in which no discrete state appears more than once except for the first and the last discrete states is a cycle. The sequence of discrete states CVS 9, CVS 6, CVS 8, CVS 9 is a cycle. We can find all types of cycles with the graph theory [40]. In this book, suppose that the total number of the types of cycles is θ_0 and these cycles are denoted as $LC(1), LC(2), \cdots, LC(\theta_0)$. For simplicity, assume that cycle $LC(j)$ is of the form $v_1, v_2, \cdots, v_{k_0(j)}, v_1$, where $v_i \in \{$ CVS $j, j = 1, 2, \cdots, n \}$. There is only one cycle in a quasi-periodic switched and impulsive system. For simplicity, cycle $LC(1)$ is assumed to be of the form $1, 2, \cdots, n, 1$.

Example 2.2.1. Consider a switched system composed of four CVSs where CVSs 1 and 2 are stable and CVSs 3 and 4 are unstable. The switchings of CVSs are illustrated in Fig. 2.3.

There are totally four cycles listed as follows:

LC(1): CVS 1, CVS 3, CVS 1;

LC(2): CVS 2, CVS 4, CVS 2;

LC(3): CVS 1, CVS 3, CVS 2, CVS 4, CVS 1;

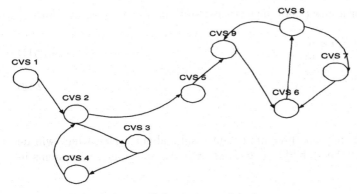

Fig. 2.2. An illustration of the concept of cycle

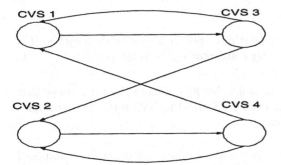

Fig. 2.3. The switchings of CVSs

LC(4): CVS 2, CVS 4, CVS 1, CVS 3, CVS 2.

This example represents the enumeration of cycles when the design of switched systems is investigated. □

Two features of the operations along cycle $LC(j)$ are given below.

1. v_{i+1} is the CVS that is next to CVS v_i, i.e.

$$v_{i+1} = \begin{cases} v_{i+1}; & i < k_0(j) \\ v_1; & i = k_0(j) \end{cases} . \tag{2.9}$$

2. v_{i-1} is the CVS that is followed by CVS v_i, i.e.

$$v_{i-1} = \begin{cases} v_{i-1}; & i > 1 \\ v_{k_0(j)}; & i = 1 \end{cases} . \tag{2.10}$$

Similarly, for a quasi-periodic switched and impulsive system, we have

$$i + 1 = \begin{cases} i+1; & i < n \\ 1; & i = n \end{cases},$$

(2.11)

$$i - 1 = \begin{cases} i-1; & i > 1 \\ n; & i = 1 \end{cases}.$$

(2.12)

We shall now present three important results about cycles, which will be used throughout the stability analysis of switched and impulsive systems in this book.

Lemma 2.2.1. *Any closed path can be decomposed into a non–empty set of cycles.*   □

Lemma 2.2.2. *There exist at least one cycle in a path if its length is $(l + 1)(l = 1, 2, \cdots)$ and there are only l different CVSs in the path.*   □

For any two time instant t_0 and t, let $\aleph_{t_0, t}$ denote the total times that each cycle appears in the path from CVS $m(t_0)$ to CVS $m(t)$. Two examples are given below for illustration:

Example 2.2.2. Consider a switched and impulsive system that is composed of CVSs 1 and 2. Clearly, the total number of cycle is 1, $\aleph_{t_{s,1}^1, t_{s,1}^{k+1}} = k$, and $\aleph_{t_{s,1}^1, t_{s,2}^k} = k - 1$.   □

Example 2.2.3. Consider the switched system that is given in Example 2.2.1. Suppose that the path from $m(t_0)$ to $m(t)$ is CVS 1, CVS 3, CVS 1, CVS 3, CVS 1, CVS 3, CVS 1, CVS 3, CVS 2, CVS 4, CVS 2, CVS 4, CVS 2, CVS 4, CVS 1, CVS 3, CVS 1. Then $\aleph_{t_0, t}$ is 6.   □

It is known from Lemma 2.2.1 that

$$\aleph_{t_{s,i}^k, t_{s,i}^{k+1}} \geq 1,$$

(2.13)

$$\aleph_{t_0, t_{s,i}^{k+1}} \geq k.$$

(2.14)

Note that the number of CVSs is finite. Clearly, there exists a CVS that appears infinite number of times in a switched and impulsive system with infinite number of switchings. We then have the following lemma.

Lemma 2.2.3. $\aleph_{t_0, t} \to \infty$ *as* $t \to \infty.$ □

Lemmas 2.2.1-2.2.3 can be used to derive less conservative conditions on the stability analysis of switched and impulsive systems. It will be shown that the stability of switched and impulsive systems is totally determined by the behavior of the system along each type of cycle. Thus, we shall focus our analysis on each type of cycle and the method is called the *cycle analysis method*. With the method, a new notion, redundancy of each cycle, is introduced to relax the conditions on the stability of switched and impulsive systems. Since the total number of cycles is finite, it is very easy to verify the results derived in this book. This is very important for the application of switched and impulsive systems.

The *cycle analysis method* studies switched and impulsive systems in a unified way. Each cycle, instead of each CVS, is chosen as a unit for the stability analysis. The redundancy of each cycle is thus removed. This is the major reason why less conservative and easier-to-check conditions can be derived.

Although the *cycle analysis method* is useful to study the stability of switched and impulsive systems, some cycles are divided into small parts with each part embedded into a different part of the discrete trajectory of the system. For example, consider a switched system as shown in Fig. 2.4. There are three cycles listed as

LC(1): CVS 1 → CVS 2 → CVS 4 → CVS 3 → CVS 1,
LC(2): CVS 4 → CVS 5 → CVS 7 → CVS 4,
LC(3): CVS 4 → CVS 5 → CVS 6 → CVS 7 → CVS 4.

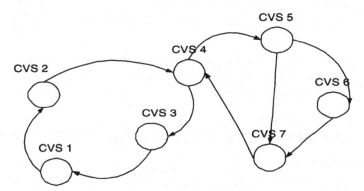

Fig. 2.4. A simple example of switched system

Suppose that the initial CVS is CVS 1. A possible discrete trajectory of the above example within some interval could be CVS 1, CVS 2, CVS 4, CVS 5, CVS 6, CVS 4, CVS 5, CVS 7, CVS 4, CVS 3, where CVS 3 is the ending point. The corresponding continuous trajectory is given as

$$X(t_{s,1}^{i+1}) = e^{A(3)\Delta\tau_3} e^{A(4)\Delta\tau_4} e^{A(7)\Delta\tau_7} e^{A(5)\Delta\tau_5} e^{A(4)\Delta\tau_4}$$
$$e^{A(6)\Delta\tau_6} e^{A(5)\Delta\tau_5} e^{A(4)\Delta\tau_4} e^{A(2)\Delta\tau_2} e^{A(1)\Delta\tau_1} X(t_{s,1}^i).$$

This is a closed path that contains cycles 1-3. However, cycle 1 is divided into two parts. One part of it is embedded at the beginning while the other part is embedded at the end of the trajectory. To use the *cycle analysis method*, the state evolution of each cycle should be transformed into a scalar quantity so that commutative manipulations can be performed. As it is, this cannot be achieved directly because $AB = BA$ does not always hold. To overcome this, there are two possible approaches. One is to find a map that converts a matrix and a vector into a scalar, and then combine each type of cycle together. Lyapunov functions, matrix measures and generalized matrix measures will be applied to combine each type of cycle together, and at the same time it removes the redundancy of each cycle in the analysis. The other way is to impose some constraints such that each cycle is not divided.

2.3 The State Space of Switched and Impulsive Systems

To formulate the stability theory of switched and impulsive systems mathematically, we shall first define its metric space.

The state space of switched and impulsive systems is denoted as $\bar{S}$. Let $\rho(S(t_1), S(t_2))$ denote the distance from $S(t_1)$ to $S(t_2)$ and

$$\rho(S(t_1), S(t_2)) = \rho(X(t_1), X(t_2)) + \rho(m(t_1), m(t_2)), \tag{2.15}$$

where

$$\rho(X(t_1), X(t_2)) = \sqrt{(X(t_1) - X(t_2))^T (X(t_1) - X(t_2))}, \tag{2.16}$$

$$\rho(m(t_1), m(t_2)) = \begin{cases} 1 & \text{if } m(t_1) \neq m(t_2) \\ 0 & \text{if } m(t_1) = m(t_2) \end{cases}. \tag{2.17}$$

When the number of CVSs is 1, equation (2.15) becomes

$$\rho(S(t_1), S(t_2)) = \rho(X(t_1), X(t_2)). \tag{2.18}$$

Furthermore, the distance from the state $S(t)$ to the state subset $\widetilde{S}$ is defined by

$$\rho(S(t), \widetilde{S}) = \inf_{S(t') \in \widetilde{S}} \{\rho(S(t), S(t'))\}. \tag{2.19}$$

Hence $(\bar{S}, \rho)$ is a metric space.

2.4 Basic Definitions on Lyapunov Stability

The stability of a physical system generally results from the presence of mechanisms which dissipate energy. Stability is a very old research topic, dating back to the advent of theory for differential equation (the original paper by Maxwell). The objective of stability analysis is to draw conclusions about the behavior of a system without actually computing its solution trajectory. It is also the key point for a system because only a stable system can work well. If a system is not stable, then it may burn out (in the case of electrical systems), disintegrate (in the case of mechanical systems), overflow (in the case of computer systems), or congestion (in the case of networks). Thus, an unstable system is useless. Roughly speaking, a stable system is the one in which a small input will yield a response that does not diverge.

Definition 2.4.1. *Define $REA(S(t))$ as the set of states that are reachable from $S(t)$. $\bar{S}_M$ is an invariant set if for all $S(t) \in \bar{S}_M$, we have $REA(S(t)) \subseteq \bar{S}_M$. $\bar{S}_M$ is an equilibrium if for all $S(t) \in \bar{S}_M$, we have $X(t) = X_e$.* □

Example 2.4.1. Consider Example 1.2.1 again. Suppose that $X_1(t_0)+X_2(t_0)+X_3(t_0) = 1$. An invariant set for the example is

$$\{(m(t), X(t))|m(t) \text{is in an invariant set},$$
$$X_1(t) \geq 0, X_2(t) \geq 0, X_3(t) \geq 0; X_1(t) + X_2(t) + X_3(t) = 1 > 0\}. \tag{2.20}$$

A subset of the set (2.20) is

$$\bar{S}_M =$$
$$\{m(t) = 1, 0 \leq X_1(t) \leq \frac{2}{3}, \sum_{i=1}^{3} X_i(t) = 1, X_2(t) = X_3(t) + \frac{1}{3}\}$$
$$\cup \{m(t) = 2, 0 \leq X_1(t) \leq \frac{1}{3}, \sum_{i=1}^{3} X_i(t) = 1, X_3(t) = X_1(t) + \frac{1}{3}\}$$
$$\cup \{m(t) = 3, \frac{1}{3} \leq X_1(t) \leq \frac{2}{3}, \sum_{i=1}^{3} X_i(t) = 1, X_1(t) = X_2(t) + \frac{1}{3}\}. \tag{2.21}$$

This is illustrated in Fig. 2.5. □

With the concepts of an equilibrium and an invariance set, the stability of switched and impulsive systems can de defined as follows:

Definition 2.4.2. *A switched and impulsive system is said to be*

- *stable in the sense of Lyapunov with respect to an invariant set $\bar{S}_M$ if for any $\xi > 0$ there exists a quantity $\delta(\xi, t_0) > 0$ such that when $\rho(S(t_0), \bar{S}_M) < \delta$ we have $\rho(S(t), \bar{S}_M) < \xi$ for all $t \geq t_0$.*

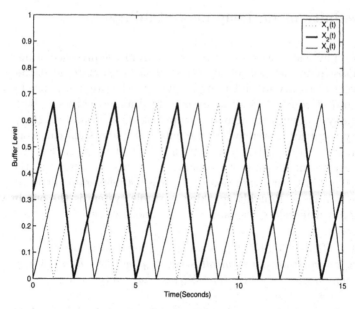

Fig. 2.5. An invariant set of Example 1.2.1

- *globally stable in the sense of Lyapunov with respect to $\bar{S}_M$ if $\delta = \infty$.*

- *uniformly stable in the sense of Lyapunov with respect to $\bar{S}_M$ if $\delta = \delta(\xi)$.*

- *asymptotically stable in the sense of Lyapunov with respect to $\bar{S}_M$ if it is stable and $\rho(S(t), \bar{S}_M) \to 0$ as $t \to +\infty$.*

- *uniformly asymptotically stable in the sense of Lyapunov with respect to $\bar{S}_M$ if it is asymptotically stable in the sense of Lyapunov and for each $\xi > 0$ there exists a $\bar{t} = t(\xi) > 0$ such that $\rho(S(t), \bar{S}_M) < \xi$ for all $t \geq \bar{t}$.*

- *exponentially stable in the sense of Lyapunov with respect to $\bar{S}_M$ if it is asymptotically stable in the sense of Lyapunov with respect to $\bar{S}_M$ and there exist a positive number M and a negative number λ such that*

$$\rho(S(t), \bar{S}_M) \leq M e^{\lambda(t-t_0)} \rho(S(t_0), \bar{S}_M). \tag{2.22}$$

□

Definition 2.4.3. *Assume that there exist a function $V(m(t), X(t)) : \{1, 2, \cdots, n\} \times R^r \to R^+$ that is in a predefined set $\emptyset$. Then the system*

$$\begin{cases} w(t) \in \emptyset \\ w(t_0) = w_0 \geq 0 \end{cases} \tag{2.23}$$

is the comparison system of switched and impulsive system (1.8) and (1.10).

□

2.5 Two Basic Theorems on Lyapunov Stability

The Lyapunov theory is a very popular and effective approach for studying the stability of switched and impulsive systems. The idea behind our method is that if we have a Lyapunov function for each CVS, then we only need to put restrictions on the switchings of the system to guarantee the stability of switched and impulsive system. Based on this observation, the following result is derived on the uniform stability of switched and impulsive systems.

Theorem 2.5.1. *A switched and impulsive system (1.8) and (1.10) is uniformly stable in the sense of Lyapunov with respect to $\bar{S}_M$ if there exist Lyapunov functions $V(m(t), X(t))$ such that the following conditions hold:*

Condition I *There exist two positive numbers c_1 and c_2 to bound the Lyapunov functions as*

$$c_1 \rho^2(S(t), \bar{S}_M) \le V(m(t), X(t)) \le c_2 \rho^2(S(t), \bar{S}_M). \tag{2.24}$$

Condition II *There exist $\omega_i \in C[R^+, R^+]$ with $\omega_i(0) = 0$ to satisfy that*

$$\sup_{t_{s,i}^k < t \le t_{f,i}^k} V(i, X(t)) \le \omega_i(V(i, X(t_{s,i}^k))), \tag{2.25}$$

where $i = 1, 2, \cdots, n, k = 1, 2, \cdots, +\infty$.

Condition III *There exists a $\nu \in C[R^+, R^+]$ with $\nu(0) = 0$ such that*

$$V(i, X(t_{s,i}^k)) \le \nu(V(m(t_0), X(t_0))), \tag{2.26}$$

where $i = 1, 2, \cdots, n, k = 1, 2, \cdots, +\infty$.

Proof. It can be shown that the comparison system of switched and impulsive system (1.8) and (1.10) is

$$\begin{cases} \sup_{t_{s,i}^k < t \le t_{f,i}^k} w(t) \le \omega_i(w(t_{s,i}^k)) \\ w(t_{s,i}^k) \le \nu(w(t_0^+)) \\ w(t_0) = w_0 \ge 0 \end{cases} \tag{2.27}$$

Since $\nu(0) = 0$ and $\omega_i(0) = 0$, it can be proven that the system (2.27) is uniformly stable. Clearly, the original switched and impulsive system (1.8) and (1.10) is uniformly stable in the sense of Lyapunov with respect to $\bar{S}_M$.   □

The result on the uniform asymptotic stability can be further developed as in the following theorem.

Theorem 2.5.2. *A switched and impulsive system (1.8) and (1.10) is uniformly asymptotically stable in the sense of Lyapunov with respect to $\bar{S}_M$ if there exist Lyapunov functions $V(m(t), X(t))$ such that the Conditions I and II of Theorem 2.5.1 hold and the following conditions hold:*

Condition IV *Lyapunov functions are decreasing along a subsequence of the switchings, i.e. there exist an $L(0 < L < 1)$ and a subsequence of the switchings $\{t_{i_k}, k = 0, 1, 2, \cdots\}$ such that*

$$V(m(t_{i_{k+1}}), X(t_{i_{k+1}})) < L^2 V(m(t_{i_k}), X(t_{i_k})). \tag{2.28}$$

Condition V *There exists a $\Phi \in C[R^+, R^+]$ with $\Phi(0) = 0$ such that for any $j, i_k < j < i_{k+1} (k = 0, 1, \cdots)$,*

$$V(m(t_j), X(t_j)) \leq \Phi(V(m(t_{i_k}), X(t_{i_k}))). \tag{2.29}$$

Condition VI *For any $j, 0 \leq j \leq i_0$, there exists a $\Psi \in C[R^+, R^+]$ with $\Psi(0) = 0$ to satisfy that*

$$V(m(t_j), X(t_j)) \leq \Psi(V(m(t_0), X(t_0))). \tag{2.30}$$

Proof. It can be shown that the comparison system of switched and impulsive system (1.8) and (1.10) is

$$\begin{cases} \sup_{t_{s,i}^k < t \leq t_{f,i}^k} w(t) \leq \omega_i(w(t_{s,i}^k)) \\ w(t_{i_{k+1}}) \leq Lw(t_{i_k}) \\ w(t_j) \leq \Phi(w(t_{i_k})) \; ; \; i_k < j < i_{k+1} \\ w(t_j) \leq \Psi(w(t_0)) \; ; \; 0 \leq j \leq i_0 \\ w(t_0) = w_0 \geq 0 \end{cases} \tag{2.31}$$

It can be easily proven that the system (2.31) is uniformly asymptotically stable. Clearly, the original switched and impulsive system (1.8) and (1.10) is uniformly asymptotically stable in the sense of Lyapunov with respect to $\bar{S}_M$. $\qquad\square$

Remark 2.5.1. If the selected Lyapunov function satisfies Condition I, Condition VI is then also a necessary condition for the uniform asymptotic stability of switched and impulsive systems in the sense of Lyapunov with respect to $\bar{S}_M$. It can be shown as follows:

From the definition, we have

$$\rho(S(t_k), \bar{S}_M) = 0 \quad as \quad k \to \infty.$$

It follows that

$$V(m(t_k), X(t_k)) \to 0 \quad as \quad k \to \infty.$$

For $\xi = \vartheta_5 V(m(t_0), X(t_0))$, *there exists* t_{i_1}, *such that*

$$V(m(t_{i_1}), X(t_{i_1})) \le \vartheta_5 V(m(t_0), X(t_0)).$$

Similarly, we have a sequence $\{t_{i_k}\}$ *which satisfies that*

$$V(m(t_{i_{k+1}}), X(t_{i_{k+1}})) \le \vartheta_5 V(m(t_{i_k}), X(t_{i_k})),$$

where $k = 1, 2, \cdots$ □

Remark 2.5.2. Theorems 2.5.1 and 2.5.2 are very crucial and useful results because Lyapunov functions are only required to be non-increasing along a subsequence of the switchings while other existing results require Lyapunov functions to be non-increasing along the whole switching sequence [128]. With our result, it is possible to introduce a new notion, i.e. redundancy of each cycle, to study the stability of switched and impulsive systems. However, so far, no general procedure has been provided to identify the non-increasing subsequence, and the continuous functions ω_i, Ψ *and* Φ.   □

2.6 An Illustrative Example

In this section, we shall use the results derived in the previous section to study the stability of Example 1.2.1 with respect to an invariant set in the sense of Lyapunov.

Example 2.6.1. Consider Example 1.2.1. Suppose that $X_1(t_0) + X_2(t_0) + X_3(t_0) = 1$. Then the system is uniformly asymptotically stable with respect to the invariant set (2.21). The is illustrated in Figs. 2.6 and 2.7.

To verify the result via the application of Theorems 2.5.1 and 2.5.2, we choose the Lyapunov functions as

$$
\begin{aligned}
&V(m(t), X(t)) \\
&= \rho^2(S(t), \bar{S}_M) \\
&= \begin{cases}
(X_1(t) - \frac{2}{3})^2 + (X_2(t) - \frac{1}{3})^2 + X_3^2(t); & X_1(t) > \frac{2}{3}, m(t) = 1 \\
(\frac{X_1(t)}{2} + X_2(t) - \frac{2}{3})^2 + (\frac{X_1(t)}{2} + X_3(t) - \frac{1}{3})^2; & 0 \le X_1(t) \le \frac{2}{3}, m(t) = 1 \\
X_1^2(t) + (X_2(t) - \frac{2}{3})^2 + (X_3(t) - \frac{1}{3})^2; & X_2(t) > \frac{2}{3}, m(t) = 2 \\
(\frac{X_2(t)}{2} + X_3(t) - \frac{2}{3})^2 + (X_1(t) + \frac{X_2(t)}{2} - \frac{1}{3})^2; & 0 \le X_2(t) \le \frac{2}{3}, m(t) = 2 \\
(X_1(t) - \frac{1}{3})^2 + X_2^2(t) + (X_3(t) - \frac{2}{3})^2; & X_3(t) > \frac{2}{3}, m(t) = 3 \\
(X_1(t) + \frac{X_3(t)}{2} - \frac{2}{3})^2 + (X_2(t) + \frac{X_3(t)}{2} - \frac{1}{3})^2; & 0 \le X_3(t) \le \frac{2}{3}, m(t) = 3
\end{cases}
\end{aligned}
$$

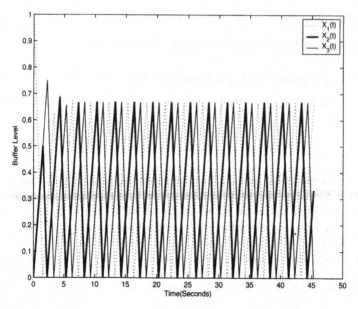

Fig. 2.6. The state evolution with initial value $[1, 0, 0]^T$

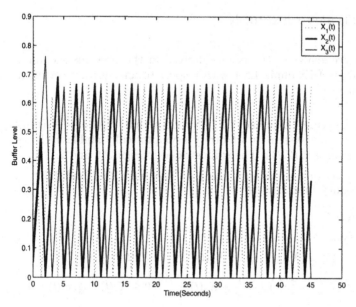

Fig. 2.7. The state evolution with initial value $[0.85, 0.05, 0.1]^T$

Define $D^+V(m(t), X(t))$ as [86]

$$D^+V(m(t), X(t)) = D^+V(i, X(t))$$
$$= \lim_{\delta \to 0^+} \sup \frac{1}{\delta}[V(i, X(t)) + \delta f(X(t), i) - V(i, X(t))],$$

where $t \in [t_{s,i}^k, t_{f,i}^k] (k = 1, 2, \cdots, i = 1, 2, \cdots, n)$.

All the conditions of Theorems 2.5.1 and 2.5.2 are verified below.

1. Condition II of Theorem 2.5.1 is shown as follows:

   Note that

   $$D^+V(1, X(t)) = \begin{cases} \frac{2}{3}(2 - 3X_1(t)); & X_1(t) \geq \frac{2}{3} \\ 0; & X_1(t) \leq \frac{2}{3} \end{cases},$$

   $$D^+V(2, X(t)) = \begin{cases} \frac{2}{3}(2 - 3X_2(t)); & X_2(t) \geq \frac{2}{3} \\ 0; & X_2(t) \leq \frac{2}{3} \end{cases},$$

   $$D^+V(3, X(t)) = \begin{cases} \frac{2}{3}(2 - 3X_3(t)); & X_3(t) \geq \frac{2}{3} \\ 0; & X_3(t) \leq \frac{2}{3} \end{cases}.$$

   It follows that

   $$D^+V(1, X(t)) \leq 0,$$
   $$D^+V(2, X(t)) \leq 0,$$
   $$D^+V(3, X(t)) \leq 0.$$

   Thus, we have
   $$\omega_1(X) = \omega_2(X) = \omega_3(X) = X.$$

2. The non-increasing subsequence of switchings is identified by

   Proposition 2.6.1. *$V(m(t), X(t))$ is non-increasing along subsequence* $\{t_3, t_9, t_{15}, \cdots, t_{6k+3}, \cdots\}$. *Furthermore,*

   $$V(m(t_{6k+9}), X(t_{6k+9})) \leq (\frac{1}{64})^2 V(m(t_{6k+3}), X(t_{6k+3})).$$

   Proof. The CVS of the switched system is CVS 1 within $[t_{6k}, t_{6k+1}] \cup [t_{6k+3}, t_{6k+4}]$. By the initial continuous states, the state equations and the reset maps, we get

$$X_1(t_{6k+3}) + X_2(t_{6k+3}) = 1,$$
$$X_3(t_{6k+3}) = 0,$$
$$X^T(t_{6k+4}) = \left[0 \ \frac{X_1(t_{6k+3})}{2} + X_2(t_{6k+3}) \ \frac{X_1(t_{6k+3})}{2} \right],$$
$$X^T(t_{6k+5}) = \left[\frac{X_1(t_{6k+3})}{4} + \frac{X_2(t_{6k+3})}{2} \ 0 \ \frac{3X_1(t_{6k+3})}{4} + \frac{X_2(t_{6k+3})}{2} \right],$$
$$X^T(t_{6k+6}) = \left[\frac{5X_1(t_{6k+3})}{8} + \frac{3X_2(t_{6k+3})}{4} \ \frac{3X_1(t_{6k+3})}{8} + \frac{X_2(t_{6k+3})}{4} \ 0 \right],$$
$$X^T(t_{6k+7}) = \left[0 \ \frac{11X_1(t_{6k+3})}{16} + \frac{5X_2(t_{6k+3})}{8} \ \frac{5X_1(t_{6k+3})}{16} + \frac{3X_2(t_{6k+3})}{8} \right],$$
$$X^T(t_{6k+8}) = \left[\frac{11X_1(t_{6k+3})}{32} + \frac{5X_2(t_{6k+3})}{16} \ 0 \ \frac{21X_1(t_{6k+3})}{32} + \frac{11X_2(t_{6k+3})}{16} \right],$$
$$X^T(t_{6k+9}) = \left[\frac{43X_1(t_{6k+3})}{64} + \frac{21X_2(t_{6k+3})}{32} \ \frac{21X_1(t_{6k+3})}{64} + \frac{11X_2(t_{6k+3})}{32} \ 0 \right].$$

Consider the following two cases:

a) When $X_1(t_{6k+3}) > \frac{2}{3}$, we have

$$V(m(t_{6k+3}), X(t_{6k+3})) = 2(X_1(t_{6k+3}) - \frac{2}{3})^2,$$
$$V(m(t_{6k+9}), X(t_{6k+9})) = (\frac{1}{64})^2 \cdot 2(X_1(t_{6k+3}) - \frac{2}{3})^2.$$

They imply

$$V(m(t_{6k+9}), X(t_{6k+9})) = (\frac{1}{64})^2 (V(m(t_{6k+3}), X(t_{6k+3}))). \qquad (2.32)$$

b) When $a \le \frac{2}{3}$,

$$V(m(t_{6k+3}), X(t_{6k+3})) = 2(\frac{X_1(t_{6k+3})}{2} - \frac{1}{3})^2,$$
$$V(m(t_{6k+9}), X(t_{6k+9})) = (\frac{1}{64})^2 \cdot 2(\frac{X_1(t_{6k+3})}{2} - \frac{1}{3})^2.$$

They yield

$$V(m(t_{6k+9}), X(t_{6k+9})) = (\frac{1}{64})^2 (V(m(t_{6k+3}), X(t_{6k+3}))). \qquad (2.33)$$

It follows from equations (2.32) and (2.33) that

$$V(m(t_{6k+9}), X(t_{6k+9})) \le (\frac{1}{64})^2 (V(m(t_{6k+3}), X(t_{6k+3}))).$$

$\square$

3. Conditions V and VI of Theorem 2.5.2 are verified by the following results:

Proposition 2.6.2. *For $j = 1, 2, 3, 4, 5$,*
$$V(m(t_{6k+3+j}), X(t_{6k+3+j})) \le V(m(t_{6k+3}), X(t_{6k+3})). \qquad (2.34)$$

$\square$

Proposition 2.6.3. *For $j = 1, 2, 3$,*

$$V(m(t_j), X(t_j)) \leq V(m(t_0), X(t_0)).$$ (2.35)

□

By Theorems 2.5.1 and 2.5.2, the switched server system is uniformly asymptotically stable in the sense of Lyapunov with respect to $\bar{S}_M$ defined in (2.21). □

Remark 2.6.1. Comparing with the Clear–the–Largest–Buffer–Level method [155], our simple cyclic control policies 1 and 2 have two advantages.

1. They are very simple.

2. There exists only one limit cycle.

□

Remark 2.6.2. The stability results for client/server systems in the general case can be found in [123, 122]. The scheduling method provided in this example is called a simple cyclic control policy. It was shown in [123, 122] that the scheduling method can be used together with other existing methods, like the Clear–the–Largest–Buffer–Level method, to design new scheduling methods with pretty good performance.

For example, the switching law can be set as follows:

When the buffer i is empty, it switches to a buffer j with $X_i(j) \geq X_i(t)(\forall i \neq j)$.

It is an optimal and stable real time scheduling method that can be applied to a flexible manufacturing system [122]. □

3. Stability of Linear Switched and Impulsive Systems

In the previous chapter, we have set up the basic framework for the stability analysis of switched and impulsive systems. The framework provides the possibility of obtaining better stability results.

With Theorems 2.5.1 and 2.5.2 and Lemmas 2.2.1 and 2.2.2, a new notion of redundancy of each cycle can be introduced to study the stability of linear switched and impulsive systems. With this, each cycle, instead of each CVS, is treated as a unit for the stability analysis.

The similar idea has already been used in the information theory to improve the source coding gain. The essence of the source coding is to use short symbols representing values that occur frequently, and long symbols representing values that occur less frequently. To further improve the coding gain, each successive N samples are treated as a vector-sample and assign a codeword for it [41].

In the information theory, the redundancy is always measured by entropy. However in this chapter, the redundancy of each cycle will be measured and removed by three types of tools, namely, multiple Lyapunov functions, matrix norms and matrix measures. Less conservative conditions will then be derived on the stability of the linear switched and impulsive systems

$$\begin{cases} \dot{X}(t) = A(m(t))X(t) \; ; \; t \neq t_j \\ X(t_j^+) = D(m(t_j^-), m(t_j^+))X(t_j^-) \; . \\ m(t_j^+) \in \psi(m(t_i^-)) \end{cases} \tag{3.1}$$

We denote CVS i as mode i of a linear switched and impulsive system. Then for CVS i, all the corresponding functions can be written as $A(i)$, $D(i,j)$, $\psi(i)$ etc.

Multiple Lyapunov functions and matrix measures are applicable for a wider class of linear switched and impulsive systems while matrix norms can only be used to study the stability of a special class of linear switched and impulsive systems.

3.1 Multiple Lyapunov Functions

In this section, we shall use multiple Lyapunov functions to analyze the stability of linear switched and impulsive systems. To help readers understand better the development of the stability analysis, we shall start from the simple case of quasi–periodic switched and impulsive systems, where there is only one cycle in the system. A typical example is the temperature adjustment system within our body during the four seasons: spring, summer, autumn and winter. Besides the adaptivity of our body, different clothes have to be worn to protect and help our body adapt to different seasons.

Before presenting the result, basic knowledge on symmetric and definite matrices is given. Let A be a square matrix. As usual, we say that A is symmetric if $A = A^T$. The matrix A (not necessarily symmetric) is said to be nonnegative definite if $v^T A v \geq 0$ for all vectors v. We say that A is positive definite if A is nonnegative definitive and $v^T A v$ implies $v = 0$. Sometimes, we write $A \geq 0 (A > 0)$ meaning that A is nonnegative (positive) definitive.

Theorem 3.1.1. *A switched and impulsive system of the form (3.1) is uniformly asymptotically stable in the sense of Lyapunov with respect to $X_e = 0$ if there exist positive definite matrices $P_1, \cdots, P_n$ such that the following two conditions hold.*

Condition 1 : *There exist $a_i \in R$ and $b_i \in R$ such that*

$$\kappa_i I > A^T(i)P_i + P_i A(i) ; \quad i = 1, 2, \cdots, n, \qquad (3.2)$$
$$a_i P_i > D^T(i, i+1)P_{i+1}D(i, i+1) ; \quad i = 1, 2, \cdots, n. \qquad (3.3)$$

Condition 2: *Lyapunov functions are decreasing along the cycle $1, 2, \cdots, n, 1$, i.e.*

$$G(1) \triangleq \frac{\ln(\prod_{i=1}^n a_i \prod_{i \in \Gamma_2} e^{\frac{\kappa_i}{\lambda_{max}(P_i)} \Delta_{1,i}} \prod_{i \in \Gamma_1} e^{\frac{\kappa_i}{\lambda_{min}(P_i)} \Delta_{2,i}})}{2} < 0, \qquad (3.4)$$

where

$$\Gamma_1 = \{i | \kappa_i > 0\}, \Gamma_2 = \{i | \kappa_i \leq 0\}.$$

Proof. Define the Lyapunov functions as

$$V(X(t), m(t)) = X^T(t)P(m(t))X(t),$$

where

$$P(m(t)) = \begin{cases} P_1; & \text{CVS } 1 \\ \vdots & \vdots \\ P_n; & \text{CVS } n \end{cases}.$$

To use Theorem 2.5.2, we need to prove that (3.5), (3.6) and (3.7) hold. That is

1. There exists a $\omega_i \in C[R^+, R^+]$ with $\omega_i(0) = 0$ such that

$$\sup_{t_{s,i}^k < t \leq t_{f,i}^k} V(X(t), i) \leq \omega_i(V(X(t_{s,i}^k), i)), \tag{3.5}$$

where

$$\omega_i(z) = \begin{cases} e^{\frac{\kappa_i}{\lambda_{min}(P_i)} \Delta_{2,i}} z; & i \in \Gamma_1 \\ z; & i \in \Gamma_2 \end{cases}.$$

2. We have

$$V(X(t_{s,1}^{k+1}), 1) \leq L^2 V(X(t_{s,1}^k), 1). \tag{3.6}$$

where $L = e^{G(1)}$.

3. For any $i (i \in \{1, 2, \cdots, n\})$,

$$V(X(t_{s,i}^k), i) \leq \hbar_1^{2n-2} \hbar_2^2 L^{2N_{t_{s,1}^k, t_{s,i}^k}} V(X(t_{s,1}^k), 1), \tag{3.7}$$

where

$$\hbar_1 = \max\{1, \max_i\{\sqrt{a_i}\}\}, \tag{3.8}$$

$$\hbar_2 = \prod_{i \in \Gamma_1} e^{\frac{\kappa_i}{2\lambda_{min}(P_i)} \Delta_{2,i}}. \tag{3.9}$$

This means that $\Phi(z) = \Psi(z) = \hbar_1^{2n-2} \hbar_2^2 L^{2N_{t_{s,1}^k, t_{s,i}^k}} z$.

These conditions are now proved in detail as follows.

Note that

$$\lambda_{max}(P_i)\|X(t)\|^2 \geq V(X(t), i) \geq \lambda_{min}(P_i)\|X(t)\|^2. \tag{3.10}$$

From inequalities (3.2) and (3.10), we have

$$\frac{\partial V(X(t), i)}{\partial X(t)} A(i) X(t) \leq \kappa_i \|X(t)\|^2$$

$$\leq \begin{cases} \frac{\kappa_i}{\lambda_{min}(P_i)} V(X(t), i); & i \in \Gamma_1 \\ \frac{\kappa_i}{\lambda_{max}(P_i)} V(X(t), i); & i \in \Gamma_2 \end{cases}.$$

It follows that

$$V(X(t_{f,i}^k), i) \leq \begin{cases} e^{\frac{\kappa_i}{\lambda_{min}(P_i)} (t_{f,i}^k - t_{s,i}^k)} V(X(t_{s,i}^k), i); & i \in \Gamma_1 \\ e^{\frac{\kappa_i}{\lambda_{max}(P_i)} (t_{f,i}^k - t_{s,i}^k)} V(X(t_{s,i}^k), i); & i \in \Gamma_2 \end{cases}.$$

Thus, we obtain

a) For $i \in \Gamma_1$,

$$V(X(t_{f,i}^k), i) \leq e^{\frac{\kappa_i}{\lambda_{min}(P_i)} \Delta_{2,i}} V(X(t_{s,i}^k), i), \tag{3.11}$$

$$V(X(t), i) \leq e^{\frac{\kappa_i}{\lambda_{min}(P_i)} \Delta_{2,i}} V(X(t_{s,i}^k), i) \; ; \; t_{s,i}^k < t < t_{f,i}^k. \tag{3.12}$$

b) For $i \in \Gamma_2$,

$$V(X(t_{f,i}^k), i) \le e^{\frac{\kappa_i}{\lambda_{max}(P_i)} \Delta_{1,i}} V(X(t_{s,i}^k), i), \tag{3.13}$$

$$V(X(t), i) \le V(X(t_{s,i}^k), i) \; ; \; t_{s,i}^k < t < t_{f,i}^k. \tag{3.14}$$

From inequalities (3.13), (3.14), (3.11) and (3.12) and the conditions of the theorem, we can show that inequalities (3.5) holds.

From condition (3.3), we can get

$$V(X(t_{s,j+1}^k), j+1) \le a_i V(X(t_{f,j}^k), j). \tag{3.15}$$

We derive from (3.13) and (3.15) that for any t, $(t_{s,i}^k \le t \le t_{f,i}^k)$,

$$
\begin{aligned}
V(X(t), i) &\le \max\{1, e^{\frac{\kappa_i}{\lambda_{min}(P_i)} \Lambda_3(i)}\} V(X(t_{s,i}^k), i) \\
&\le a_{i-1} \max\{1, e^{\frac{\kappa_i}{\lambda_{min}(P_i)} \Lambda_3(i)}\} V(X(t_{f,i-1}^k), i-1) \\
&\le a_{i-1} \max\{1, e^{\frac{\kappa_i}{\lambda_{min}(P_i)} \Lambda_3(i)}\} e^{\frac{\kappa_{i-1}}{\lambda_{min}(P_{i-1})} \Lambda_3(i-1)} V(X(t_{s,i-1}^k), i-1) \\
&\le \cdots \\
&\le \prod_{j=1}^{i-1} a_j \max\{1, e^{\frac{\kappa_i}{\lambda_{min}(P_i)} \Lambda_3(i)}\} \prod_{j=1}^{i-1} e^{\frac{\kappa_j}{\lambda_{min}(P_j)} \Lambda_3(j)} V(X(t_{s,1}^k), 1),
\end{aligned}
$$

where

$$\Lambda_3(i) = \begin{cases} \Delta_{2,i}; \; i \in \Gamma_1 \\ \Delta_{1,i}; \; i \in \Gamma_2 \end{cases}.$$

Let $i = 1$ and $t = t_{f,1}^{k+1}$, we obtain inequality (3.6).

Note also that

$$\prod_{j=1}^{i-1} a_j \max\{1, e^{\frac{\kappa_i}{\lambda_{min}(P_i)} \Lambda_3(i)}\} \prod_{j=1}^{i-1} e^{\frac{\kappa_j}{\lambda_{min}(P_j)} \Lambda_3(j)} \le \hbar_1^{2n-2} \hbar_2^2 L^{2\aleph_{t_{s,1}^k, t_{s,i}^k}}$$

holds for any i ($i \in \{1, 2, \cdots, n\}$). It follows that inequality (3.7) holds.

Using Theorem 2.5.2, the linear switched and impulsive system of form (3.1) is uniformly asymptotically stable in the sense of Lyapunov with respect to $X_e = 0$. $\square$

Remark 3.1.1. *It is interesting to note that the discrete trajectory of a quasi-periodic linear switched and impulsive system forms a cycle of the form 1, 2, $\cdots$, n, 1 and repeats within the cycle. In the above theorem, our result is derived by considering the system behavior along the cycle. Note also that there are only a finite number of cycles in a general switched and impulsive system [98]. Thus, similar stability conditions can be obtained by considering the behavior of the general switched and impulsive systems along each type of cycle.* $\square$

Based on the above proof process, the stability of the linear switched and impulsive systems is studied from the following three perspectives:

1. **Continuous perspective**

   Study the Lyapunov function along each CVS to measure its redundancy.

2. **Discrete perspective**

   Analyze the logical behavior to identify each type of cycle, which is set as the unit for the stability analysis.

3. **Hybrid perspective**

   Consider the Lyapunov functions along each type of cycle to remove the redundancy of each cycle.

Moreover, it is known from the above remarks and results that a linear switched and impulsive system is stable if there exist multiple Lyapunov functions which are non–increasing along each type of cycle. Since the total number of cycle is finite, the conditions can be easily verified in practice as illustrated in the following numerical example.

Example 3.1.1. Consider a quasi–periodic linear switched and impulsive system of the form (3.1), which is composed of two CVSs given as follows:

$$\text{CVS 1:} \quad A(1) = \begin{bmatrix} 1/6 & 0 \\ 0 & 1/3 \end{bmatrix},$$

$$\text{CVS 2:} \quad A(2) = \begin{bmatrix} -1/2 & 0 \\ 0 & -3/4 \end{bmatrix},$$

$$\text{Impulsive switching 1:} \quad D(1,2) = \begin{bmatrix} 1 & 0 \\ 0 & 2 \end{bmatrix},$$

$$\text{Impulsive switching 2:} \quad D(2,1) = \begin{bmatrix} 1/4 & 0 \\ 0 & 1/4 \end{bmatrix}.$$

The system will automatically switch from CVS 1 to CVS 2 or CVS 2 to CVS 1, when the duration time of CVS 1 or CVS 2 reaches 2 or 6 units, respectively.

The matrices P_1 and P_2 are chosen as

$$P_1 = \begin{bmatrix} 2 & 0 \\ 0 & 1 \end{bmatrix}, P_2 = \begin{bmatrix} 1 & 0 \\ 0 & 2/3 \end{bmatrix}.$$

It can be easily computed that

$$a_1 = 4, a_2 = 1, \kappa_1 = 3/4, \kappa_2 = -0.9,$$
$$a_1 a_2 e^{2\kappa_1} e^{6\kappa_2} = 3.e^{-3.9} < 1.$$

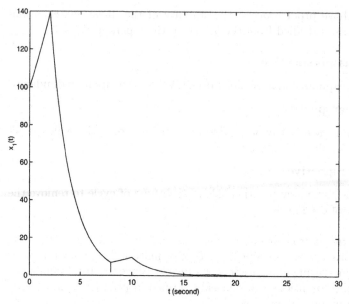

Fig. 3.1. The time responses $X_1(t)$ of the system

Therefore, the quasi–periodic switched and impulsive system is uniformly stable in the sense of Lyapunov. To illustrate the result, let the initial state of the system be $X_1(t_0) = 100$ and $X_2(t_0) = 100$. The trajectories of the system in this case are illustrated as Figs. 3.1 and 3.2.   □

Theorem 3.1.1 is then applied to study the stabilization of the bilinear systems via switched control. Consider the following bilinear system [32]:

$$\dot{X}(t) = AX(t) + by(t)u(t), y(t) = cX(t), \tag{3.16}$$

where $u(t)$ is a single control input, (A, c) is observable, (A, b) is controllable, and $cb = 0$.

The switched control

$$u = \begin{cases} -\frac{1}{y}kX; & \text{if } |y| > \epsilon\|X\| \\ 0; & \text{otherwise} \end{cases} \tag{3.17}$$

is designed for the bilinear system (3.16).

We divide the state space into two sets Ω^- and Ω^+ with a boundary Ω^o as

$$\Omega^- = \{X||y| \le \epsilon\|X\|\}, \tag{3.18}$$
$$\Omega^+ = \{X||y| > \epsilon\|X\|\}, \tag{3.19}$$
$$\Omega^o = \{X||y| = \epsilon\|X\|\}. \tag{3.20}$$

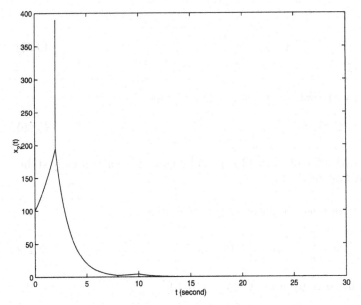

Fig. 3.2. The time responses $X_2(t)$ of the system

Let $\{t_i\}$ be a non-increasing time subsequence, where t_{2i}'s denote the time instants when the state $X(t)$ exits Ω^+ to enter Ω^-, and t_{2i+1}'s the time instants when $X(t)$ exits Ω^- to enter Ω^+.

The resultant closed-loop system can be described by a linear switched system

$$\text{CVS 1: } \dot{X}(t) = (A - bK)X(t), \tag{3.21}$$
$$\text{CVS 2: } \dot{X}(t) = AX(t). \tag{3.22}$$

The switching law is as defined equations (3.18)-(3.20). Obviously, there exists only one cycle: CVS 1 → CVS 2 → CVS 1. It is thus a quasi-periodic switched systems.

We choose the Lyapunv function as $V(X(t)) = X^T(t)PX(t)$, where

$$(A - bK)^T P + P(A - bK) = -Q < 0. \tag{3.23}$$

We can then easily conclude that the following results hold [32].

1. For any $t \in [t_{2i}, t_{2i+1}]$,

$$V(t) \le e^{\psi(t-t_{2i})}V(t_{2i}), \tag{3.24}$$

where $\psi = \bar{\delta}_{\tilde{Q}}/\underline{\delta}_P$, $\tilde{Q} = A^T P + PA$, $\underline{\delta}_P$ is the minimum singular value of P, and $\bar{\delta}_{\tilde{Q}}$ is the maximum singular value of $\tilde{Q}$.

2. For any $t \in [t_{2i+1}, t_{2i+2}]$,

$$V(t) \leq e^{-\gamma(t-t_{2i})} V(t_{2i}),$$ (3.25)

where $\gamma = \bar{\delta}_Q / \underline{\delta}_P$.

3. For any i, there exists an $a(0 < a < 1)$ such that

$$V(t_{2i+2}) \leq aV(t_{2i}).$$ (3.26)

Therefore, the division controller (3.17) can be used to asymptotically stabilize the bilinear system (3.16).

Example 3.1.2. Consider a bilinear system as follows:

$$A = \begin{bmatrix} 3 & 0 \\ 0 & -2 \end{bmatrix}, b = \begin{bmatrix} 1 \\ -1 \end{bmatrix}, c = \begin{bmatrix} 1 & 1 \end{bmatrix}.$$ (3.27)

The initial condition is given by $X^T(0) = [10, 10]$.

Choose $\epsilon = 0.1$ and $K = [3.25, 1.25]$. It can be easily shown that the eigenvalues of $(A - bK)$ is $-0.5 \pm 2j$. The switched controller

$$u = \begin{cases} -\frac{1}{y}3.25X_1 + 1.25X_2; & \text{if } |y| > 0.1\|X\| \\ 0; & \text{otherwise} \end{cases}$$ (3.28)

can be used to stabilize the bilinear system (3.27). □

3.2 Matrix Measures

Although multiple Lyapunov functions are applicable for a wider class of linear switched and impulsive systems, it is very difficult to find the Lyapunov functions $(P_1, P_2, \cdots, P_n)$ to satisfy (3.2)–(3.4), especially when there are some unstable CVSs. In this section, we introduce another tool, i.e. matrix measures, to remove the redundancy of each cycle.

The matrix measure, $\mu(A)$, has the following properties [188]:

$$\mu(A + B) = \mu(A) + \mu(B) \; ; \; \forall A, B \in R^{r \times r},$$ (3.29)

$$\|e^{At}\| \leq e^{\mu(A)t} \; ; \; \forall A \in R^{r \times r} \; ; \; \forall t \geq 0,$$ (3.30)

$$\mu(A) \leq \|A\| \; ; \; \forall A \in R^{r \times r},$$ (3.31)

$$\mu(aA) = a\mu(A) \; ; \; \forall a \geq 0 \; ; \; A \in R^{r \times r}.$$ (3.32)

Sufficient conditions are derived on the stability of switched linear systems (3.1) by using the *cycle analysis method* and matrix measure.

For simplicity, we first study the case that the duration time of CVS $i(i = 1, 2, \cdots, n)$ is fixed as $\Delta\tau_i$. Consider the behavior of linear switched and impulsive system along cycle $LC(j)$ in the form of $v_1, v_2, \cdots, v_{k(j)}, v_1$. Then we have

$$X(t_{s,v_1}^{k+1}) = \prod_{l=1}^{k_0(j)} (D(v_l, v_{l+1})e^{A(v_l)\Delta\tau_{v_l}})X(t_{s,v_1}^k). \qquad (3.33)$$

Denote

$$G(j) = \sum_{l=1}^{k_0(j)} [\ln \|D(v_l, v_{l+1})\| + \mu(A(v_l))\Delta\tau_{v_l}] \; ; \; j = 1, 2, \cdots, \theta_0, \qquad (3.34)$$

$$\Upsilon(j) = \prod_{l=1}^{k_0(j)} (D(v_l, v_{l+1})e^{A(v_l)\Delta\tau_{v_l}}) \; ; \; j = 1, 2, \cdots, \theta_0, \qquad (3.35)$$

$$\Gamma_1 = \{i|\mu(A(i)) \geq 0\}, \qquad (3.36)$$

$$\Gamma_2 = \{i|\mu(A(i)) < 0\}. \qquad (3.37)$$

Then, we obtain the following result:

Theorem 3.2.1. *A linear switched and impulsive system of the form (3.1) with a fixed duration time for each CVS is uniformly asymptotically stable in the sense of Lyapunov with respect to $X_e = 0$ if $G(j)$ defined in (3.34) is less than 0 for all $j \in \{1, 2, \cdots, \theta_0\}$.*

Proof. Define the Lyapunov function as

$$V(X(t), m(t)) = X^T(t)X(t) = \|X(t)\|^2.$$

It follows that

$$\sup_{t_{s,i}^k \leq t < t_{f,i}^k} V(X(t), i) \leq \omega_i(V(X(t_{s,i}^k), i)). \qquad (3.38)$$

where

$$\omega_i(z) = \begin{cases} e^{2\mu(A(i))\Delta_{2,i}}z; & i \in \Gamma_1 \\ z; & i \in \Gamma_2 \end{cases}.$$

Since there are infinite number of switchings and the number of CVSs is n, there exists at least a CVS which appears infinite number of times. Assume that CVS i_0 is the first one that appears infinite number of times. Then, we have

$$V(X(t_{s,i_0}^{k+1}), i_0) \leq \mathrm{L}^{2\aleph_{t_{s,i_0}^k, t_{s,i_0}^{k+1}}} V(X(t_{s,i_0}^k), i_0) \; ; \; k = 2, 3, 4, \cdots \qquad (3.39)$$

where $\mathrm{L} = \max_{1 \leq j \leq \theta_0}\{e^{G(j)}\}$.

Let $\hbar_1 = \max\{1, \max_{i,j} \|D(i,j)\|\}$ and $\hbar_2 = \prod_{i \in \Gamma_1} e^{\mu(A(i))\Delta\tau_i}$. Then we get

$$V(X(t_l^+), m(t_l^+)) \leq \hbar_1^{2n-2} \hbar_2^2 \mathrm{L}^{2\aleph_{t_{s,i_0}^k, t_l^+}} V(X(t_{s,i_0}^k), i_0) \tag{3.40}$$

for any switching instant t_l between t_{s,i_0}^{k+1} and t_{s,i_0}^k. In other words, $\Phi(z) = \hbar_1^{2n-2} \hbar_2^2 \mathrm{L}^{2\aleph_{t_{s,i_0}^k, t_l^+}} z$.

Similarly, using Lemmas 2.2.1 and 2.2.2, we get

$$V(X(t_l^+), m(t_l^+)) < \hbar_1^{2n-2} \hbar_2^2 \mathrm{L}^{2\aleph_{t_0, t_l^+}} V(X(t_0), m(t_0)). \tag{3.41}$$

holds for any switching instant t_l between t_{s,i_0}^1 and t_0. This implies that $\Psi(z) = \hbar_1^{2n-2} \hbar_2^2 \mathrm{L}^{2\aleph_{t_0, t_l^+}} z$.

From inequalities (3.38), (3.39), (3.40), (3.41) and Theorem 2.5.2, we know that Theorem 3.2.1 holds. □

Consider the case where $D(i,j) = I(i, j \in \{1, 2, \cdots, n\})$, that is, there is no impulsive switchings. $G(j)$ in equation (3.34) becomes

$$G(j) = \sum_{l=1}^{k_0(j)} \mu(A(v_l))\Delta\tau_{v_l} \ ; \ j = 1, 2, \cdots, \theta_0. \tag{3.42}$$

From Theorem 3.2.1, we can obtain the following corollary:

Corollary 3.2.1. *A linear switched system of the form (3.1) is uniformly asymptotically stable in the sense of Lyapunov with respect to $X_e = 0$ if $G(j)$ defined in (3.42) is less than 0 for all $j \in \{1, 2, \cdots, \theta_0\}$.* □

Remark 3.2.1. If each type of cycle in the non-impulsive switching system is composed of all $i(i \in \{1, 2, \cdots, n\})$, i.e. each type of cycle is of the form $i_1, i_2, \cdots, i_n, i_1 (i_j, i_k \in M, i_j \neq i_k$ for $j \neq k)$, then Theorem 1 in [49] is derived. That is, Theorem 1 in [49] is a special case of Corollary 3.2.1. □

For linear switched and impulsive systems in the case that the duration time of each CVS is in an interval, we denote

$$G(j) = \sum_{l=1}^{k_0(j)} [\ln \|D(v_l, v_{l+1})\| + \mu(A(v_l))\hat{\epsilon}(v_l)], \tag{3.43}$$

where

$$\hat{\epsilon}(i) = \begin{cases} \Delta_{2,i} & i \in \Gamma_1 \\ \Delta_{1,i} & i \in \Gamma_2 \end{cases}. \tag{3.44}$$

Similarly, we have the following stability result:

Theorem 3.2.2. *A linear switched and impulsive system of the form (3.1) is uniformly asymptotically stable in the sense of Lyapunov with respect to $X_e = 0$ if $G(j)$ defined in (3.43) is less than 0 for all $j \in \{1, 2, \cdots, \theta_0\}$.* □

When $D(i,j) = I(i,j \in \{1, 2, \cdots, n\})$, i.e. there is no impulsive switching, $G(j)$ in equation (3.43) becomes

$$G(j) = \sum_{l=1}^{k(j)} \mu(A(v_l))\hat{\epsilon}(v_l). \tag{3.45}$$

We then have

Corollary 3.2.2. *A linear switched system of the form (3.1) is uniformly asymptotically stable in the sense of Lyapunov with respect to $X_e = 0$ if $G(j)$ defined in (3.45) is less than 0 for all $j \in \{1, 2, \cdots, \theta_0\}$.* □

It can be readily seen from the results obtained in this and the previous sections, that both the Lyapunov function and the matrix measure convert a matrix quantity into a scalar. Since scalars are always commutative, sufficient conditions like (3.4), (3.34), (3.42), (3.43) and (3.45) were derived to verify the stability of linear switched and impulsive systems. Also from the above examples, these conditions can be verified without any difficulty.

3.3 Matrix Norms

Whenever we use multiple Lyapunov functions or matrix measures, there must exist at least one stable CVS within the overall system. However, there do exist linear switched and impulsive systems which are totally composed of unstable CVSs. In this section, matrix norms will be applied to study the stability of such linear switched and impulsive systems.

Note that some cycles may be divided into several small parts and each one is embedded into a different part of the discrete trajectory. Thus, an assumption should be imposed such that there exists at most one cycle which is separated at most once.

Assumption 3.1 *There exists only one common CVS i_0 in each cycle, i.e. for any cycles $LC(j_1)$ and $LC(j_2)$*

$$LC(j_1) \cap LC(j_2) = \{i_0\} \; ; \; j_1, j_2 \in \{1, 2, \cdots, \theta_0\}; j_1 \neq j_2.$$

□

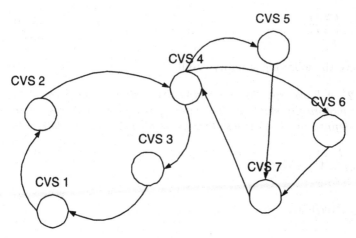

Fig. 3.3. Illustration of Assumption 3.1

The above assumption is illustrated in Fig. 3.3, where there are three cycles and CVS 4 is the common model. With Assumption 3.1, we then have the following result.

Theorem 3.3.1. *Suppose that linear switched and impulsive systems (3.1) satisfy Assumption 3.1. They are uniformly asymptotically stable in the sense of Lyapunov with respect to $X_e = 0$ if all $\Upsilon(j)$ $(j \in \{1, 2, \cdots, \theta_0\})$ that are given in equation (3.35) satisfy*

$$L \stackrel{\Delta}{=} \max_{1 \le j \le \theta_0} \|\Upsilon(j)\| < 1. \tag{3.46}$$

$\square$

A necessary and sufficient condition can be obtained for a limited class of switched and impulsive systems, as stated below.

Theorem 3.3.2. *The linear switched and impulsive systems of the form (3.1) are uniformly asymptotically stable in the sense of Lyapunov with respect to $X_e = 0$ if and only if there exists a matrix norm $\| \cdot \|_d$ such that*

$$L \stackrel{\Delta}{=} \max_{1 \le j \le \theta_0} \|\Upsilon(j)\|_d < 1. \tag{3.47}$$

$\square$

Remark 3.3.1. Although matrix norms can only be used to study a limited class of linear switched and impulsive systems, they can be used to derive some necessary and sufficient conditions. Note that Lyapunov functions and matrix measures can only be used to derive some sufficient conditions. $\square$

In the next numerical example, the matrix norm is applied to study the stability of linear switched and impulsive systems.

Example 3.3.1. Consider a linear switched and impulsive system of the form (3.1), which is composed of two CVSs. The models and reset maps are

$$\text{CVS 1:}\ \ A(1) = \begin{bmatrix} 1/4 & 0 \\ 0 & -1/2 \end{bmatrix},$$

$$\text{CVS 2:}\ \ A(2) = \begin{bmatrix} -1 & 0 \\ 0 & 0.25 \end{bmatrix},$$

$$\text{Impulsive switching 1:}\ \ D(1,2) = \begin{bmatrix} 1 & 0 \\ 0 & 0.5 \end{bmatrix},$$

$$\text{Impulsive switching 2:}\ \ D(2,1) = \begin{bmatrix} 1/2 & 0 \\ 0 & 1 \end{bmatrix}.$$

The CVS will automatically switch from CVS 1 to CVS 2 and CVS 2 to CVS 1, when the duration time of CVS 1 or CVS 2 reaches 2 units, respectively.

Note that there is only one cycle in this example and

$$\Upsilon(1) = \begin{bmatrix} \frac{e^{-1.5}}{2} & 0 \\ 0 & \frac{e^{-0.5}}{2} \end{bmatrix}.$$

Obviously,

$$\|\Upsilon(1)\| < 1.$$

Using Theorem 3.3.1, we know that the linear switched and impulsive system given in this example is stable. Note that the two individual CVSs are unstable.   □

3.4 A Decomposition Method

For those linear switched and impulsive systems that are totally composed of unstable CVSs, but do not satisfy Assumption 3.1, the results in the previous sections cannot be used to study their stability. In this case, we introduce a state transformation to decompose the original system into several lower dimensional systems. Every lower dimensional system shall consist of both stable and unstable systems and the stability of such lower dimensional systems can then be studied by using the results obtained.

The class of linear switched and impulsive systems considered in this section is composed of a finite number of CVSs:

$$\dot{X}(t) = A(m(t))X(t), \tag{3.48}$$

where each $A(i)$ is unstable.

At time t_j, impulsive "switchings" will occur as follows [10]:

$$\begin{cases} X(t_j^+) = D(m(t_j^-), m(t_j^+))X(t_j^-) \\ m(t_j^+) \in \psi(m(t_j^-)) \end{cases}, \tag{3.49}$$

where each $D(m(t_j^-), m(t_j^+))$ is also unstable.

To present the decomposition method, we shall introduce the following definition:

Definition 3.4.1. *A state transformation $\tilde{X} = TX$ is said to be a decomposition mapping of linear switched and impulsive systems (3.48) and (3.49) if it satisfies the following two conditions:*

1. T is nonsingular.

2. The following two equations hold.

$$TAT^{-1} = \begin{bmatrix} \tilde{A}_{11}(m(t)) & \tilde{A}_{12}(m(t)) & \cdots & \tilde{A}_{1r_0}(m(t)) \\ 0 & \tilde{A}_{22}(m(t)) & \cdots & \tilde{A}_{2r_0}(m(t)) \\ \vdots & \vdots & \ddots & \vdots \\ 0 & 0 & \cdots & \tilde{A}_{r_0r_0}(m(t)) \end{bmatrix}$$

$$TDT^{-1} = \begin{bmatrix} \tilde{D}_{11}(m(t_j^-), m(t_j^+)) & \tilde{D}_{12}(m(t_j^-), m(t_j^+)) & \cdots & \tilde{D}_{1r_0}(m(t_j^-), m(t_j^+)) \\ 0 & \tilde{D}_{22}(m(t_j^-), m(t_j^+)) & \cdots & \tilde{D}_{2r_0}(m(t_j^-), m(t_j^+)) \\ \vdots & \vdots & \ddots & \vdots \\ 0 & 0 & \cdots & \tilde{D}_{r_0r_0}(m(t_j^-), m(t_j^+)) \end{bmatrix}$$

where $\tilde{D}_{ll}(m(t_j^-), m(t_j^+)) \in R^{r_l \times r_l}$, $\tilde{A}_{ll}(m(t)) \in R^{r_l \times r_l}$, $m(t_j^-), m(t_j^+), m(t) \in \{1, 2, \cdots, n\}$, $l \in \{1, 2, \cdots, r_0\}$, $\sum_{i=1}^{r_0} = r$ and $r_0 \in \{1, 2, \cdots, r\}$.

Assume that there exists such a decomposition mapping $\tilde{X} = TX$ for linear switched and impulsive systems (3.48) and (3.49). Then, a state transformation can be defined as

$$\tilde{X} = [\tilde{X}_1, \tilde{X}_2, \cdots, \tilde{X}_{r_0}]^T = TX.$$

Under the transformation, we have

$$\begin{bmatrix} \dot{\tilde{X}}_1(t) \\ \dot{\tilde{X}}_2(t) \\ \vdots \\ \dot{\tilde{X}}_{r_0}(t) \end{bmatrix} = \begin{bmatrix} \tilde{A}_{11}(m(t)) & \tilde{A}_{12}(m(t)) & \cdots & \tilde{A}_{1r_0}(m(t)) \\ 0 & \tilde{A}_{22}(m(t)) & \cdots & \tilde{A}_{2r_0}(m(t)) \\ \vdots & \vdots & \ddots & \vdots \\ 0 & 0 & \cdots & \tilde{A}_{r_0r_0}(m(t)) \end{bmatrix} \begin{bmatrix} \tilde{X}_1(t) \\ \tilde{X}_2(t) \\ \vdots \\ \tilde{X}_{r_0}(t) \end{bmatrix}, \tag{3.50}$$

and

$$\begin{bmatrix} \tilde{X}_1(t_j^+) \\ \tilde{X}_2(t_j^+) \\ \vdots \\ \tilde{X}_{r_0}(t_j^+) \end{bmatrix} = \begin{bmatrix} \tilde{D}_{11} & \tilde{D}_{12} & \cdots & \tilde{D}_{1r_0} \\ 0 & \tilde{D}_{22} & \cdots & \tilde{D}_{2r_0} \\ \vdots & \vdots & \ddots & \vdots \\ 0 & 0 & \cdots & \tilde{D}_{r_0r_0} \end{bmatrix} \begin{bmatrix} \tilde{X}_1(t_j^-) \\ \tilde{X}_2(t_j^-) \\ \vdots \\ \tilde{X}_{r_0}(t_j^-) \end{bmatrix}. \tag{3.51}$$

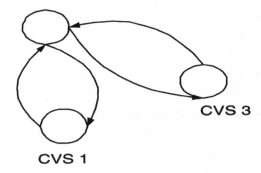

CVS 2

CVS 3

CVS 1

Fig. 3.4. The switchings of a linear switched system

Example 3.4.1. Consider a linear switched system that is composed of three CVSs, where the switchings are given as in Fig. 3.4. The three system matrices are

$$A(1) = \begin{bmatrix} 2.5 & 7 & 5 & 2 \\ -3.75 & -2.75 & 1.25 & -1.75 \\ 2.8333 & 1.9167 & -1.75 & 0.75 \\ -1.5833 & -1.1667 & 0.5 & -2 \end{bmatrix},$$

$$A(2) = \begin{bmatrix} -3 & 1 & 7 & 1 \\ 5.25 & 3.5 & -1.5 & 0 \\ -4.5833 & -6 & -2 & 0.5 \\ -1.6667 & -7.5 & -12.5 & -4.5 \end{bmatrix},$$

$$A(3) = \begin{bmatrix} 3 & 7 & 3 & 0 \\ -5 & -8 & -2.5 & 0.5 \\ 5.333 & 8.5 & 3.6667 & 0.1667 \\ -5.3333 & -14.5 & -11.1667 & -1.6667 \end{bmatrix}.$$

The duration times of CVS 1, 2 and 3 are within $[3, 5], [4.5, 5.5]$ and $[4, 6]$, respectively.

Note that $A(1)$, $A(2)$ and $A(3)$ are all unstable. Thus, the stability of the linear switched and impulsive system cannot be studied directly by using the results in the previous sections. Let

$$T = \begin{bmatrix} 1 & 0 & 0 & 0 \\ 2 & 2 & 0 & 0 \\ 0.5 & 3 & 3 & 0 \\ 1 & 1 & 1 & 1 \end{bmatrix}.$$

Then, we have

$$TA(1)T^{-1} = \begin{bmatrix} -2 & 1 & 1 & 2 \\ -1 & 2 & 4 & 0.5 \\ 0 & 0 & 1 & -2 \\ 0 & 0 & 2 & -1 \end{bmatrix},$$

$$TA(2)T^{-1} = \begin{bmatrix} 1 & -3 & 2 & 1 \\ 3 & -1 & 3 & 2 \\ 0 & 0 & -3 & 2 \\ 0 & 0 & -2 & -3 \end{bmatrix},$$

$$TA(3)T^{-1} = \begin{bmatrix} -1.5 & 2 & 1 & 0 \\ -2 & -1.5 & 0 & 1 \\ 0 & 0 & 1 & 2 \\ 0 & 0 & -2 & -1 \end{bmatrix}.$$

We can obtain the following two lower dimensional systems:

$$\dot{\hat{X}}_1(t) = \tilde{A}_{11}(m(t))\hat{X}_1(t),$$
$$\dot{\hat{X}}_2(t) = \tilde{A}_{22}(m(t))\hat{X}_2(t),$$

where

$$\tilde{A}_{11}(1) = \begin{bmatrix} -2 & 1 \\ -1 & -2 \end{bmatrix} ; \ \tilde{A}_{11}(2) = \begin{bmatrix} 1 & -3 \\ 3 & -1 \end{bmatrix} ; \ \tilde{A}_{11}(3) = \begin{bmatrix} -1.5 & 2 \\ -2 & -1.5 \end{bmatrix} ;$$
$$\tilde{A}_{22}(1) = \begin{bmatrix} 1 & -2 \\ 2 & -1 \end{bmatrix} ; \ \tilde{A}_{22}(2) = \begin{bmatrix} -3 & 2 \\ -2 & -3 \end{bmatrix} ; \ \tilde{A}_{22}(3) = \begin{bmatrix} 1 & 2 \\ -2 & -1 \end{bmatrix}.$$

$\square$

Theorem 3.4.1. *The linear switched and impulsive systems (3.48) and (3.49) are asymptotically stable if and only if the corresponding linear switched and impulsive systems (3.50) and (3.51) are asymptotically stable.*

To use the results obtained in the previous three sections, we also need the following definition:

Definition 3.4.2. *A linear switched and impulsive system (3.48) and (3.49) is said to belong to Class LSS if there exists a decomposition mapping $\tilde{X} = TX$ such that there exists at least an $\tilde{A}_{ll}(i)(i \in \{1, 2, \cdots, n\})$ that is stable for each $l(l = 1, 2, \cdots, r_0)$ in the corresponding transformed linear switched and impulsive system (3.50) and (3.51).*

For a system belonging to Class LSS, we have the following result.

Theorem 3.4.2. *Consider a linear switched and impulsive system (3.50) and (3.51) transformed from equations (3.48) and (3.49) and belonging to Class LSS. It is asymptotically stable in the sense of Lyapunov with respect to $X_e = 0$ if the lower dimensional system*

$$\begin{cases} \dot{\hat{X}}_l(t) = \tilde{A}_{ll}(m(t))\hat{X}_l(t) \\ \hat{X}_l(t_j^+) = \tilde{D}_{ll}(m(t_j^-), m(t_j^+))\hat{X}_l(t_j^-) \; ; \; l = 1, 2, \cdots, r_0 \end{cases} \tag{3.52}$$

satisfies the conditions of Theorem 3.2.1.

Proof. For simplicity, we only consider a simple case in which the linear switched and impulsive system is composed of two CVSs and $D(1, 2) = D(2, 1) = I$. Two CVSs are given as follows:

$$\dot{X}(t) = A(1)X(t), \tag{3.53}$$

$$\dot{X}(t) = A(2)X(t), \tag{3.54}$$

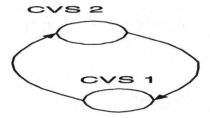

Fig. 3.5. A simple switched linear system

and the switchings of the system are illustrated in Fig. 3.5. The general case can be proven in a similar way. From Definition 3.4.2, we know that there exists a T such that

$$TA(i)T = \begin{bmatrix} \tilde{A}_{11}(i) & \tilde{A}_{12}(i) \\ 0 & \tilde{A}_{22}(i) \end{bmatrix} \; ; \; i = 1, 2. \tag{3.55}$$

From the proof of Theorem 1 [102], we know that there exists an L satisfying $0 \le \text{L} < 1$ such that

$$\|e^{\tilde{A}_{11}(1)(t_{f,1}^k - t_{s,1}^k)} e^{\tilde{A}_{11}(2)(t_{f,2}^k - t_{s,2}^k)} \hat{X}_1\| \le \text{L}\|\hat{X}_1\|, \tag{3.56}$$

$$\|e^{\tilde{A}_{22}(1)(t_{f,1}^k - t_{s,1}^k)} e^{\tilde{A}_{22}(2)(t_{f,2}^k - t_{s,2}^k)} \hat{X}_2\| \le \text{L}\|\hat{X}_2\|. \tag{3.57}$$

Note that

$$\tilde{X}_1(t_{s,2}^{k+2})$$
$$= \tilde{X}_1(t_{f,1}^{k+1})$$
$$= e^{\tilde{A}_{11}(1)(t_{f,1}^{k+1}-t_{s,1}^{k+1})}\tilde{X}_1(t_{s,1}^{k+1}) + \int_{t_{s,1}^{k+1}}^{t_{f,1}^{k+1}} e^{\tilde{A}_{11}(1)(t_{f,1}^{k+1}-\tau)}\tilde{A}_{12}(1)\tilde{X}_2(t)d\tau$$
$$= e^{\tilde{A}_{11}(1)(t_{f,1}^{k+1}-t_{s,1}^{k+1})}e^{\tilde{A}_{11}(2)(t_{f,2}^{k+1}-t_{s,2}^{k+1})}\tilde{X}_1(t_{s,2}^{k+1})$$
$$+ \int_{t_{s,1}^{k+1}}^{t_{f,1}^{k+1}} e^{\tilde{A}_{11}(1)(t_{f,1}^{k+1}-\tau)}\tilde{A}_{12}(1)\tilde{X}_2(t)d\tau$$
$$+ e^{\tilde{A}_{11}(1)(t_{f,1}^{k+1}-t_{s,1}^{k+1})}\int_{t_{s,2}^{k+1}}^{t_{f,2}^{k+1}} e^{\tilde{A}_{11}(2)(t_{f,2}^{k+1}-\tau)}\tilde{A}_{12}(2)\tilde{X}_2(t)d\tau$$
$$= e^{\tilde{A}_{11}(1)(t_{f,1}^{k+1}-t_{s,1}^{k+1})}e^{\tilde{A}_{11}(2)(t_{f,2}^{k+1}-t_{s,2}^{k+1})}\tilde{X}_1(t_{s,2}^{k+1})$$
$$+ \int_{t_{s,1}^{k+1}}^{t_{f,1}^{k+1}} e^{\tilde{A}_{11}(1)(t_{f,1}^{k+1}-\tau)}\tilde{A}_{12}(1)e^{\tilde{A}_{22}(1)(t-t_{s,1}^{k+1})}e^{\tilde{A}_{22}(2)(t_{f,2}^{k+1}-t_{s,2}^{k+1})}d\tau\tilde{X}_2(t_{s,2}^{k+1})$$
$$+ e^{\tilde{A}_{11}(1)(t_{f,1}^{k+1}-t_{s,1}^{k+1})}\int_{t_{s,2}^{k+1}}^{t_{f,2}^{k+1}} e^{\tilde{A}_{11}(2)(t_{f,2}^{k+1}-\tau)}\tilde{A}_{12}(2)e^{\tilde{A}_{22}(2)(t-t_{s,2}^{k+1})}d\tau\tilde{X}_2(t_{s,2}^{k+1}),$$

and

$$\tilde{X}_2(t_{s,2}^{k+2}) = e^{\tilde{A}_{22}(1)(t_{f,1}^{k+1}-t_{s,1}^{k+1})}e^{\tilde{A}_{22}(2)(t_{f,2}^{k+1}-t_{s,2}^{k+1})}\tilde{X}_2(t_{s,2}^{k+1}).$$

Denote

$$a = \max\{\|\tilde{A}_{11}(1)\|, \|\tilde{A}_{12}(1)\|, \|\tilde{A}_{22}(1)\|, \|\tilde{A}_{11}(2)\|, \|\tilde{A}_{12}(2)\|, \|\tilde{A}_{22}(1)\|\},$$
$$\Delta T = \max\{\Delta_{2,1}, \Delta_{2,2}\}.$$

It follows that

$$\|\tilde{X}_1(t_{s,2}^{k+2})\| \le \mathrm{L}\|\tilde{X}_1(t_{s,2}^{k+1})\| + 2a\Delta T e^{2a\Delta T}\|\tilde{X}_2(t_{s,2}^{k+1})\|$$
$$\le \mathrm{L}\|\tilde{X}_1(t_{s,2}^{k+1})\| + \mathrm{L}^{k+1}2a\Delta T e^{2a\Delta T}\|\tilde{X}_2(t_{s,2}^0)\|$$
$$\le \cdots$$
$$\le \mathrm{L}^{k+2}\|\tilde{X}_1(t_{s,2}^0)\| + (k+1)\mathrm{L}^{k+1}2a\Delta T e^{2a\Delta T}\|\tilde{X}_2(t_{s,2}^0)\|,$$

and

$$\|\tilde{X}_2(t_{s,2}^{k+2}\| \le \mathrm{L}^{k+2}\|\tilde{X}_2(t_{s,2}^0)\|. \tag{3.58}$$

Similarly, it can be proven that

$$\|\tilde{X}_1(t)\| \le e^{2a\Delta T}\|\tilde{X}_1(t_{s,2}^k)\| + 2a\Delta T e^{2a\Delta T}\|\tilde{X}_2(t_{s,2}^k)\| \; ; \; t \in [t_{s,2}^k, t_{s,2}^{k+1}),$$

and

$$\|\tilde{X}_2(t)\| \le e^{2a\Delta T}\|\tilde{X}_2(t_{s,2}^k)\| \; ; \; t \in [t_{s,2}^k, t_{s,2}^{k+1}).$$ (3.59)

Note that

$$\lim_{k\to\infty} k L^k = 0.$$ (3.60)

It follows that the linear switched and impulsive system (3.48) and (3.49) is asymptotically stable. □

Remark 3.4.1. The proof of Theorem 3.4.2 can be easily generalized to a quasi–periodic linear switched and impulsive systems that are composed of n CVSs by using the fact that

$$\lim_{k\to\infty} L^k \sum_{i=1}^{k} i^m = 0,$$ (3.61)

where $0 \le L < 1$ and $m \in \{0, 1, \cdots, n-1\}$. □

Remark 3.4.2. The proof can also be extended to a general linear switched and impulsive system that is composed of a finite number of CVSs by considering the behavior of the system along each type of cycle. □

Remark 3.4.3. A very interesting problem is to verify whether a linear switched and impulsive system belongs to Class LSS. This problem has not been solved yet and we leave it as an open problem. □

Remark 3.4.4. Note that it is required that each CVS must be block triangularizable instead of triangularizable. Thus, our results are less conservative than Theorem 3 in [92]. □

A numerical example is provided here to illustrate the application of the decomposition method.

Example 3.4.2. Consider Example 3.4.1. Using Theorem 3.2.1, we know that all the two lower dimensional systems are asymptotically stable in the sense of Lyapunov. By using Theorems 3.4.1 and 3.4.2, we know that the original system is asymptotically stable in the sense of Lyapunov. □

3.5 Stability of Linear Switched and Impulsive Systems with Respect to Invariant Sets

In this section, we consider the stability of linear switched and impulsive systems modelled by

$$\dot{X}(t) = A(m(t))X(t) + b(m(t)), \tag{3.62}$$

where each $(A(i), b(i))(i \in \bar{M})$ describes a CVS with all $A(i)$ **unstable** and $b(i)$ **not necessarily all zero**. At time t_j, impulsive "switchings" will occur as

$$\begin{cases} m(t_j^+) \in \psi(m(t_j^-)) \\ X(t_j^+) = D(m(t_j^-), m(t_j^+))X(t_j^-) + e(m(t_j^-), m(t_j^+)) \end{cases} . \tag{3.63}$$

We shall derive sufficient conditions to guarantee that there exist invariant sets of the linear switched and impulsive systems (3.62) and (3.63).

Let t_{s,v_i}^k and t_{f,v_i}^k denote the kth starting time and the kth ending time of the ith CVS in the cycle $LC(j)(j = 1, 2, \cdots, \theta_0)$ of the form $v_1, v_2, \cdots, v_i, \cdots, v_{k_0(j)}, v_1$. Consider the system behavior along the cycle $LC(j)$ over $[t_{s,v_i}^k, t_{s,v_i}^{k+1}]$. From equation (3.62), we have

$$X(t_{f,v_i}^k) = e^{A(v_i)\Delta \tau_{v_i}} X(t_{s,v_i}^k)$$
$$+ \int_0^{\Delta \tau_{v_i}} e^{A(v_i)(\Delta \tau_{v_i} - t)} b(v_i) dt; \quad i = 1, 2, \cdots, k_0(j). \tag{3.64}$$

To simplify the notations and expressions, we denote

$$\tilde{A}(v_i) = e^{A(v_i)\Delta \tau_{v_i}}, \tag{3.65}$$

$$\tilde{b}(v_i) = \int_0^{\Delta \tau_{v_i}} e^{A(v_i)(\Delta \tau_{v_i} - \tau)} b(v_i) d\tau. \tag{3.66}$$

Thus, equation (3.64) can be expressed as

$$X(t_{f,v_i}^k) = \tilde{A}(v_i)X(t_{s,v_i}^k) + \tilde{b}(v_i). \tag{3.67}$$

Along the cycle $LC(j)$, the relationship between $X(t_{s,v_i}^k)$ and $X(t_{s,v_i}^{k+1})$ is

$$X(t_{s,v_i}^{k+1}) = \bar{A}(v_i, j)X(t_{s,v_i}^k) + \bar{b}(v_i, j), \tag{3.68}$$

where

$$\bar{A}(v_i, j) = \prod_{l=1}^{i-1}(D(v_l, v_{l+1})\tilde{A}(v_l)) \prod_{l=i}^{k_0(j)}(D(v_{l-1}, v_l)\tilde{A}(v_{l-1})) \tag{3.69}$$

$$\bar{b}(v_i, j) = \tilde{e}(v_i, j) + \tilde{f}(v_i, j), \tag{3.70}$$

$$\tilde{e}(v_i, j) = e(v_{i-1}, v_i) + D(v_{i-1}, v_i)\tilde{A}(v_{i-1})e(v_{i-2}, v_{i-1}) + \cdots$$

$$+ \prod_{l=1}^{i}(D(v_{l-1}, v_l)\tilde{A}(v_{l-1}))e(v_{k_0(j)-1}, v_{k_0(j)}) + \cdots$$

$$+ \prod_{l=1}^{i}(D(v_{l-1}, v_l)\tilde{A}(v_{l-1})) \prod_{l=i+2}^{k_0(j)} (D(v_{l-1}, v_l)\tilde{A}(v_{l-1}))e(v_i, v_{i+1}),$$

$$\tilde{f}(v_i, j) = D(v_{i-1}, v_i)\tilde{b}(v_{i-1}) + D(v_{i-1}, v_i)\tilde{A}(v_{i-1})D(v_{i-2}, v_{i-1})\tilde{b}(v_{i-2}) + \cdots$$

$$+ \prod_{l=1}^{i}(D(v_{l-1}, v_l)\tilde{A}(v_{l-1}))D(v_{k_0(j)-1}, v_{k_0(j)})\tilde{b}(v_{k_0(j)-1}) + \cdots$$

$$+ \prod_{l=1}^{i}(D(v_{l-1}, v_l)\tilde{A}(v_{l-1})) \prod_{l=i+2}^{k_0(j)} (D(v_{l-1}, v_l)\tilde{A}(v_{l-1}))D(v_i, v_{i+1})\tilde{b}(v_i).$$

For simplicity, we assume that $\|\bar{A}(v_i, j)\| < 1$ for all $i \in \{1, 2, \cdots, k_0(j)\}$ and $j \in \{1, 2, \cdots, \theta_0\}$, which is guaranteed by a condition stated in Theorem 3.5.1 to be given later.

To ensure that there exists an invariant set for the linear switched and impulsive system (3.62) and (3.63), we suppose that it satisfies Assumption 3.1 and the following assumption:

Assumption 3.2 *For any two cycles $LC(j)$ and $LC(l)$, if $LC(j) \cap LC(l) \neq \emptyset$, then for any CVS $i_0 \in LC(l) \cap LC(j)$, we have*

$$(I - \bar{A}(i_0, j))^{-1}\bar{b}(i_0, j) = (I - \bar{A}(i_0, l))^{-1}\bar{b}(i_0, l). \tag{3.71}$$

$\square$

Remark 3.5.1. Obviously, if $b(m(t)) = 0$ and $e(m(t_j^-), m(t_j^+)) = 0$, then Assumption 3.2 holds. In other words, the linear switched and impulsive system with an equilibrium point is a special case of this assumption. Therefore, Assumption 3.2 is less restrictive. $\square$

Remark 3.5.2. Condition (3.71) implies that the continuous state remains the same when the system switches back to CVS i_0 from other CVS in the cycles. As stated in Lemma 3.5.2 to be given later, it guarantees that the starting continuous state in CVS i_0 is fixed. $\square$

Example 3.5.1. Consider a linear switched system which consists of the following three two-dimensional CVSs:

$$\text{CVS 1: } \dot{X}(t) = \begin{bmatrix} -1 & 0 \\ -2 & 1 \end{bmatrix} X(t) + \begin{bmatrix} 2 \\ 2 \end{bmatrix},$$

$$\text{CVS 2: } \dot{X}(t) = \begin{bmatrix} 2 & 0 \\ 6 & -4 \end{bmatrix} X(t) + \begin{bmatrix} 4 \\ 4 \end{bmatrix},$$

$$\text{CVS 3: } \dot{X}(t) = \begin{bmatrix} 3 & 0 \\ 9 & -6 \end{bmatrix} X(t) + \begin{bmatrix} 6 \\ 6 \end{bmatrix}.$$

The switchings of the system are illustrated as in Fig. 3.6, which implies that Assumption 3.1 is satisfied. According to the employed switching law, we consider the case where the duration times of CVSs 1, 2 and 3 are $4s$, $1.5s$ and $1s$, respectively.

CVS 1

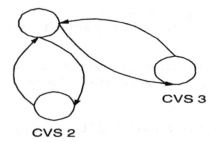

CVS 3

CVS 2

Fig. 3.6. A numerical example

It can be verified that $(I - \bar{A}(1,1))^{-1}\bar{b}(1,1) = (I - \bar{A}(1,2))^{-1}\bar{b}(1,2)$, i.e. Assumption 3.2 holds. □

To further simplify notations, we denote

$$X^e_{v_i,j} = (I - \bar{A}(v_i,j))^{-1}\bar{b}(v_i,j); \quad i = 1,2,\cdots,k_0(j), \quad j = 1,2,\cdots,\theta_0.$$

Then, for cycle $LC(j)$ $(j = 1,2,\cdots,\theta_0)$, we define

$$\bar{S}_{v_i,j} = \{(m(t),X(t))|m(t) \text{ and } X(t) \text{ are the trajectories of switched}$$
$$\text{system (3.62) and (3.63) along cycle } LC(j) \text{ with the initial state}$$
$$(m(t_0),X(t_0)) = (v_i,X^e_{v_i,j}), t_0 \leq t \leq t_0 + \sum_{l=1}^{k_0(j)} \Delta\tau_{v_l}\}. \tag{3.72}$$

Remark 3.5.3. $\bar{S}_{v_i,j}$ $(j = 1,\cdots,\theta_0)$ *are defined to determine an invariant set as stated in Lemma 3.5.3 to be given later.* □

We now prove that

1. $\bar{S}_{v_i,j}$ is not related to v_i and will be simplified as $\bar{S}_j$.

2. For any CVS v_i within a cycle $LC(j)$, the restarting continuous state is fixed at $(I - \bar{A}(v_i,j))^{-1}\bar{b}(v_i,j)$ if the first starting state is $(I - \bar{A}(v_i,j))^{-1}\bar{b}(v_i,j)$.

3. $\cup_{l=1}^{\theta_0} \bar{S}_l$ is an invariant set of the switched system (3.62) and (3.63) satisfying Assumptions 3.1 and 3.2.

To do so, we first derive the following three lemmas:

Lemma 3.5.1. *For any cycle $LC(j)(j = 1, 2, \cdots, \theta_0)$, we have*

$$X^e_{v_{i+1},j} = D(v_i, v_{i+1})\tilde{A}(v_i)X^e_{v_i,j} + D(v_i, v_{i+1})\tilde{b}(v_i) + e(v_i, v_{i+1}). \tag{3.73}$$

Proof. From (3.69), we get

$$\bar{A}(v_{i+1}, j)D(v_i, v_{i+1})\tilde{A}(v_i) = D(v_i, v_{i+1})\tilde{A}(v_i)\bar{A}(v_i, j).$$

From (3.70), we have the following equation:

$$D(v_i, v_{i+1})\tilde{A}(v_i)\bar{b}(v_i, j) + D(v_i, v_{i+1})\tilde{b}(v_i) + e(v_i, v_{i+1})$$
$$= \bar{A}(v_{i+1}, j)D(v_i, v_{i+1})\tilde{b}(v_i) + \bar{A}(v_{i+1}, j)e(v_i, v_{i+1}) + \bar{b}(v_{i+1}, j).$$

Note that

$$X^e_{v_i,j} = \bar{A}(v_i, j)X^e_{v_i,j} + \bar{b}(v_i, j).$$

It follows that

$$D(v_i, v_{i+1})\tilde{A}(v_i)X^e_{v_i,j} + D(v_i, v_{i+1})\tilde{b}(v_i) + e(v_i, v_{i+1})$$
$$= D(v_i, v_{i+1})\tilde{A}(v_i)\bar{A}(v_i, j)X^e_{v_i,j}$$
$$+ D(v_i, v_{i+1})\tilde{A}(v_i)\bar{b}(v_i, j) + D(v_i, v_{i+1})\tilde{b}(v_i) + e(v_i, v_{i+1})$$
$$= \bar{A}(v_{i+1}, j)[D(v_i, v_{i+1})\tilde{A}(v_i)X^e_{v_i,j} + D(v_i, v_{i+1})\tilde{b}(v_i)$$
$$+ e(v_i, v_{i+1})] + \bar{b}(v_{i+1}, j).$$

This implies that $D(v_i, v_{i+1})\tilde{A}(v_i)X^e_{v_i,j} + D(v_i, v_{i+1})\tilde{b}(v_i) + e(v_i, v_{i+1})$ is an equilibrium point of the equation

$$X(t^s_{v_{i+1},k+1}) = \bar{A}(v_{i+1}, j)X(t^s_{v_{i+1},k}) + \bar{b}(v_{i+1}, j). \tag{3.74}$$

Note that $\|\bar{A}(v_{i+1}, j)\| < 1$. Thus, $\bar{A}(v_{i+1}, j)$ is stable. In addition, there exists only one equilibrium point for (3.74). It follows that (3.73) holds.   □

Remark 3.5.4. *From Lemma 3.5.1, $\bar{S}_{v_i, j}$ is independent of v_i.*   □

Lemma 3.5.2. *For any CVS v_i within a cycle $LC(j)(j = 1, 2, \cdots, \theta_0)$, the re-starting continuous state is fixed at $X_{v_i, j}^e$ if the first starting continuous state is $X_{v_i, j}^e$.*   □

Lemma 3.5.3. $\bar{S}_M = \cup_{l=1}^{\theta_0} \bar{S}_l$ *is an invariant set for the linear switched and impulsive system (3.62) and (3.63) satisfying Assumptions 3.1 and 3.2.*

Proof. From Lemmas 3.5.1 and 3.5.2, and equations (3.67) and (3.68), we know that the continuous trajectories will always stay in $\bar{S}_j$ if the CVS of the system stays in cycle $LC(j)$ and the starting continuous state is in $\bar{S}_j$. Moreover, when the CVS of the system is switched from $LC(j)$ to another cycle $LC(l)$, the starting continuous state will be in $\bar{S}_l$. It follows that $\cup_{l=1}^{\theta_0} \bar{S}_l$ is an invariant set.   □

Example 3.5.2. Consider Example 3.5.1. The invariant set is

$$\bar{S}_M = \{(1, X(t)) | X_1 = X_2 = e^{-t}(1 - e^{-1})^{-1}(4e^3 - 2e^{-1} - 2)$$
$$+ 2\int_0^t e^{\tau - t} d\tau; 0 \le t \le 4\} \bigcup$$
$$\{(2, X(t)) | X_1 = X_2 = e^{2t}\left[(1 - e^{-1})^{-1}(4e^{-1} - 2e^{-5} - 2e^{-4}) + 2 - 2e^{-4}\right]$$
$$+ 4\int_0^t e^{2(t - \tau)} d\tau; 0 \le t \le 1.5\} \bigcup$$
$$\{(3, X(t)) | X_1 = X_2 = e^{3t}\left[(1 - e^{-1})^{-1}(4e^{-1} - 2e^{-5} - 2e^{-4}) + 2 - 2e^{-4}\right]$$
$$+ 6\int_0^t e^{3(t - \tau)} d\tau; 0 \le t \le 1\}.$$

It is illustrated in Fig. 3.7, where the sequence of switchings is $1 \to 2 \to 1 \to 3 \to 1 \to \cdots$.   □

In the remaining part of this section, a state transformation will be introduced to transform a linear switched and impulsive system (3.62) and (3.63) with an invariant set $\bar{S}_M$ into a linear switched and impulsive system with an equilibrium point $X_e = 0$. Theorem 3.3.1 can then be used.

Lemma 3.5.4. *A linear switched and impulsive system (3.62) and (3.63) satisfying Assumptions 3.1 and 3.2 can be converted into a linear switched and impulsive system of the form*

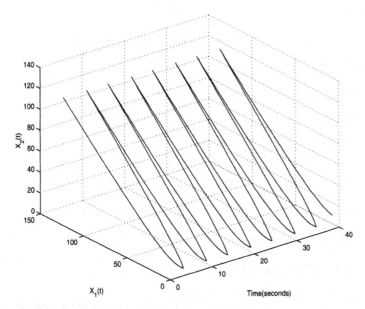

Fig. 3.7. The invariant set of Example 3.5.2

$$\dot{\hat{X}}(t) = A(v_i)\hat{X}(t), \tag{3.75}$$

$$\hat{X}(t^k_{s,v_{i+1}}) = D(v_i, v_{i+1})\hat{X}(t^k_{f,v_i}), \tag{3.76}$$

by the state transformations

$$\hat{X}(t) = X(t) - Y(t) \; ; \; t^k_{s,v_i} \le t \le t^k_{f,v_i} \; ;$$
$$i = 1, 2, \cdots, k_0(j) \; ; \; k = 1, 2, \cdots, \tag{3.77}$$

and

$$Y(t) = e^{A(v_i)(t-t^k_{s,v_i})}Y(t^k_{s,v_i}) + \int_{t^k_{s,v_i}}^{t} e^{A(v_i)(t-\tau)}b(v_i)d\tau,$$

$$Y(t^k_{s,v_i}) = X^e_{v_i,j},$$

where CVS v_i is in cycle $LC(j)(j = 1, 2, \cdots, \theta_0)$.

Proof. Note that

$$X(t) = e^{A(v_i)(t-t^k_{s,v_i})}X(t^k_{s,v_i}) + \int_{t^k_{s,v_i}}^{t} e^{A(v_i)(t-\tau)}b(v_i)d\tau.$$

It follows that

$$\hat{X}(t) = e^{A(v_i)(t - t^k_{s,v_i})} \hat{X}(t^k_{s,v_i}) \; ; \; t^k_{s,v_i} \le t \le t^k_{f,v_i} \; ; \; k = 1, 2, \cdots$$

Using Lemma 3.5.1, we have

$$
\begin{aligned}
\hat{X}(t^k_{s,v_{i+1}}) &= X(t^k_{s,v_{i+1}}) - Y(t^k_{s,v_{i+1}}) \\
&= X(t^k_{s,v_{i+1}}) - X^e_{v_{i+1},j} \\
&= D(v_i, v_{i+1}) X(t^k_{f,v_i}) + e(v_i, v_{i+1}) - X^e_{v_{i+1},j} \\
&= D(v_i, v_{i+1})(\hat{X}(t^k_{f,v_i}) + Y(t^k_{f,v_i})) + e(v_i, v_{i+1}) - X^e_{v_{i+1},j} \\
&= D(v_i, v_{i+1}) \hat{X}(t^k_{f,v_i}) + D(v_i, v_{i+1}) e^{A(v_i)\Delta \tau_{v_i}} X^e_{v_i,j} \\
&\quad + D(v_i, v_{i+1}) \int_0^{\Delta \tau_{v_i}} e^{A(v_i)(\Delta \tau_{v_i} - t)} b(v_i) dt + e(v_i, v_{i+1}) - X^e_{v_{i+1},j} \\
&= D(v_i, v_{i+1}) \hat{X}(t^k_{f,v_i}).
\end{aligned}
$$

Then, we can obtain (3.75) and (3.76). □

Remark 3.5.5. From (3.72), and Lemmas 3.5.2 and 3.5.3, we know that at any time instant t, $(m(t), Y(t)) \in \bar{S}_M$. □

Obviously, $X_e = 0$ is an equilibrium point of linear switched and impulsive system in the form of (3.75) and (3.76).

Along the cycle $LC(j)(j = 1, 2, \cdots, \theta_0)$, the relationship between $\hat{X}(t^k_{s,v_i})$ and $\hat{X}(t^{k+1}_{s,v_i})$ can be established as

$$\hat{X}(t^{k+1}_{s,v_i}) = \bar{A}(v_i, j) \hat{X}(t^k_{s,v_i}),\tag{3.78}$$

where $\bar{A}(v_i, j)$ is defined in (3.69). We now consider the stability of linear switched and impulsive system (3.75) and (3.76).

Theorem 3.5.1. *The linear switching and impulsive system of the forms (3.62) and (3.63) is uniformly asymptotically stable with respect to $\bar{S}_M$ defined in Lemma 3.5.3 if*

$$L \triangleq \max_{1 \le j \le \theta_0} \{\|\bar{A}(v_i, j)\|\} < 1.\tag{3.79}$$

Proof. We first consider the case of $m(t_0) \in \{LC(1), \cdots, LC(\theta_0)\}$. Clearly, $m(t) \in \{LC(1), \cdots, LC(\theta_0)\}$ holds for all $t \ge t_0$. Using transformation (3.77), a linear switched and impulsive system (3.62) and (3.63) with the invariant set $\bar{S}_M$ can be converted into a linear switched and impulsive system (3.75) and (3.76) with the equilibrium point $X_e = 0$. From Theorem 3.3.1, a linear switched and impulsive system (3.75) and (3.76) is uniformly asymptotically stable if $\|\bar{A}(v_i, j)\| < 1 (j = 1, \cdots, \theta_0)$. From (3.77) and Remark 3.5.5, we know that Theorem 3.5.1 holds in this case.

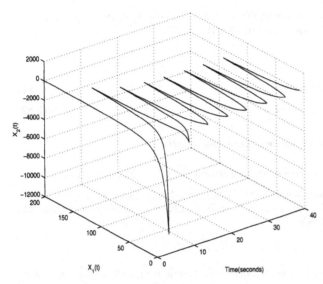

Fig. 3.8. The simulation result

Then, we consider the case that $m(t_0) \notin \{LC(1), \cdots, LC(\theta_0)\}$. According to Assumption 3.1, there exists some finite time instant t_0' such that $m(t_0') \in \{LC(1), \cdots, LC(\theta_0)\}$. In other words, the system CVS must be switched into $\{LC(1), \cdots, LC(\theta_0)\}$ after some finite time. Thus, in a similar way as above, we can conclude that Theorem 3.5.1 holds in this case.   □

Example 3.5.3. Consider Example 3.5.1 with an invariant set given in Example 3.5.2.

It can be easily verified that $\|\bar{A}(1, 1)\| = \|\bar{A}(1, 2)\| = 0.4416 < 1$.

Thus, from Theorem 3.5.1, the system is uniformly asymptotically stable with respect to the invariant set given in Example 3.5.2. This is illustrated in Fig. 3.8, where the sequence of switchings is $1 \to 2 \to 1 \to 3 \to 1 \to \cdots$.   □

3.6 Stabilization of Linear Switched Systems

Consider a linear switched system described by

$$\dot{X}(t) = A(m(t))X(t) + b(m(t))u(t), \tag{3.80}$$

where the switchings of CVSs are arbitrary, and there exists a positive dwelling time for each CVS.

This type of model can be used to represent systems subject to known abrupt parameter variations such as in communicated networks. It can also be used to approximate some types of time–varying systems. We make the following assumption about system (3.80).

Assumption 3.3

$$(A(i), b(i)) \quad \text{is controllable}, \tag{3.81}$$
$$\Delta_{1,i} = \inf_k \{t_{f,i}^k - t_{s,i}^k\} > 0, \tag{3.82}$$

where $\Delta_{1,i}$ is known and can be arbitrarily small. □

A full state feedback controller design problem can be formulated as follows:

Consider linear switched system (3.80) satisfying Assumption 3.3. Given any decay rate $\lambda < 0$, design a full state feedback controller such that the closed–loop system satisfies

$$\lim_{t \to \infty} e^{-\lambda t} \|X(t)\| = 0.$$

A switched controller is used to solve this problem and the design is mainly composed of the following two steps:

Step 1. Design a full state feedback sub-controller for each CVS.

Step 2. Define a switching law for these sub-controllers. Since the continuous state is always available, the switching instance of the system can be known exactly. Generally, the sub-controller corresponding to the active CVS should be used and the switchings of the controller should coincide exactly with those of the system. However, we cannot know the subsequent active CVSs in advance. Thus, a small delay is imposed on the switchings of the sub-controllers so that the subsequently active CVSs can be identified. Then, the sub-controller corresponding to the active CVS is switched into action.

Note that there could be overshoot in the interval before the right sub-controller is activated. Thus, the delay should be properly defined to constrain the overshoot and to properly assign the poles to ensure sufficient decay of the overshoot during the interval when the sub-controller corresponding to the active CVS is used. This is possible because the poles of each controllable CVS can be assigned arbitrarily. We shall first present two technical lemmas which will be used to assign poles for each CVS of the overall system.

Lemma 3.6.1. *For any $\lambda, \lambda_1 \in R$, let*

$$\hat{Q}_c(\lambda, \lambda_1) = \begin{bmatrix} 1 & 1 & \cdots & 1 \\ \lambda + \lambda_1 & \lambda + \lambda_1 + 1 & \cdots & \lambda + \lambda_1 + 1 \\ (\lambda + \lambda_1)^2 & (\lambda + \lambda_1)^2 + 2(\lambda + \lambda_1) & \cdots & (\lambda + \lambda_1 + 1)^2 \\ \vdots & \vdots & \ddots & \vdots \\ (\lambda + \lambda_1)^{r-1} & (\lambda + \lambda_1)^{r-1} + (r-1)(\lambda + \lambda_1)^{r-2} & \cdots & (\lambda + \lambda_1 + 1)^{r-1} \end{bmatrix},$$

$$\hat{f}_c(s) = (s - \lambda - \lambda_1)^r = s^r + d_{r-1}s^{r-1} + \cdots + d_1 s + d_0, \tag{3.83}$$

$$\hat{C}_c(\lambda, \lambda_1) = \begin{bmatrix} 0 & 1 & 0 & \cdots & 0 \\ 0 & 0 & 1 & \cdots & 0 \\ \vdots & \vdots & \vdots & \ddots & \vdots \\ -d_0 & -d_1 & -d_2 & \cdots & -d_{r-1} \end{bmatrix},$$

$$\hat{D}_c(\lambda, \lambda_1) = \begin{bmatrix} \lambda + \lambda_1 & 1 & 0 & \cdots & 0 \\ 0 & \lambda + \lambda_1 & 1 & \cdots & 0 \\ \vdots & \vdots & \vdots & \ddots & \vdots \\ 0 & 0 & 0 & \cdots & \lambda + \lambda_1 \end{bmatrix}.$$

Then

$$\hat{Q}_c^{-1}(\lambda, \lambda_1)\hat{C}_c(\lambda, \lambda_1)\hat{Q}_c(\lambda, \lambda_1) = \hat{D}_c(\lambda, \lambda_1). \tag{3.84}$$

Proof. From the definition of $\hat{f}_c(s)$ and $\hat{C}_c(\lambda, \lambda_1)$, it is clear that $\lambda + \lambda_1$ is a r-order root of polynomial $det(sI - \hat{C}_c(\lambda, \lambda_1))$. Thus

$$(\lambda + \lambda_1)^r + d_{r-1}(\lambda + \lambda_1)^{r-1} + \cdots + d_1(\lambda + \lambda_1) + d_0 = 0,$$
$$r(\lambda + \lambda_1)^{r-1} + (r-1)d_{r-1}(\lambda + \lambda_1)^{r-2} + \cdots + d_1 = 0,$$

$$\vdots$$

$$r!(\lambda + \lambda_1) + (r-1)!d_{r-1} = 0.$$

It can be shown that

$$\hat{C}_c(\lambda, \lambda_1)\hat{Q}_c(\lambda, \lambda_1)$$
$$= \begin{bmatrix} \lambda + \lambda_1 & \lambda + \lambda_1 + 1 & \cdots & \lambda + \lambda_1 + 1 \\ (\lambda + \lambda_1)^2 & (\lambda + \lambda_1)^2 + 2(\lambda + \lambda_1) & \cdots & (\lambda + \lambda_1 + 1)^2 \\ \vdots & \vdots & \ddots & \vdots \\ (\lambda + \lambda_1)^{r-1} & (\lambda + \lambda_1)^{r-1} + (r-1)(\lambda + \lambda_1)^{r-2} & \cdots & (\lambda + \lambda_1 + 1)^{r-1} \\ (\lambda + \lambda_1)^r & (\lambda + \lambda_1)^r + r(\lambda + \lambda_1)^{r-1} & \cdots & (\lambda + \lambda_1 + 1)^r \end{bmatrix},$$

and

$$\hat{Q}_c(\lambda, \lambda_1)\hat{D}_c(\lambda, \lambda_1)$$
$$= \begin{bmatrix} \lambda + \lambda_1 & \lambda + \lambda_1 + 1 & \cdots & \lambda + \lambda_1 + 1 \\ (\lambda + \lambda_1)^2 & (\lambda + \lambda_1)^2 + 2(\lambda + \lambda_1) & \cdots & (\lambda + \lambda_1 + 1)^2 \\ \vdots & \vdots & \ddots & \vdots \\ (\lambda + \lambda_1)^{r-1} & (\lambda + \lambda_1)^{r-1} + (r-1)(\lambda + \lambda_1)^{r-2} & \cdots & (\lambda + \lambda_1 + 1)^{r-1} \\ (\lambda + \lambda_1)^r & (\lambda + \lambda_1)^r + r(\lambda + \lambda_1)^{r-1} & \cdots & (\lambda + \lambda_1 + 1)^r \end{bmatrix}.$$

Therefore

$$\hat{Q}_c^{-1}(\lambda, \lambda_1)\hat{C}_c(\lambda, \lambda_1)\hat{Q}_c(\lambda, \lambda_1) = \hat{D}_c(\lambda, \lambda_1).$$

□

Lemma 3.6.2. *Let*

$$\hat{g}_i^c(s) = det(sI - A(i)) = s^r + a_{r-1}(i)s^{r-1} + \cdots + a_1(i)s + a_0(i), \qquad (3.85)$$

$$\hat{P}_c(i) = \begin{bmatrix} A^{r-1}(i)b(i) & \cdots & A(i)b(i) & b(i) \end{bmatrix} \begin{bmatrix} 1 & 0 & 0 & \cdots & 0 & 0 \\ a_{r-1}(i) & 1 & 0 & \cdots & 0 & 0 \\ \vdots & \vdots & \vdots & \ddots & \vdots \\ a_1(i) & a_2(i) & a_3(i) & \cdots & a_{r-1}(i) & 1 \end{bmatrix}.$$

If $(A(i), b(i))$ is controllable, then there exists $K(i)$, such that

$$\tilde{A}(i) = \hat{P}_c(i)\hat{Q}_c(\lambda, \lambda_1)\hat{D}_c(\lambda, \lambda_1)\hat{Q}_c^{-1}(\lambda, \lambda_1)\hat{P}_c^{-1}(i), \qquad (3.86)$$

where

$$\tilde{A}(i) = A(i) - b(i)K(i).$$

Proof. Since $(A(i), b(i))$ is controllable, it follows that $\hat{P}_c(i)$ is nonsingular. Furthermore, there exists $K(i)$ such that the closed–loop system matrix

$$\tilde{A}(i) = A(i) - b(i)K(i)$$

has all its eigenvalues equal to the sum of two arbitrary real numbers, λ and λ_1. The required $K(i)$ is

$$K(i) = \begin{bmatrix} d_0 - a_0(i) & \cdots & d_{r-1} - a_{r-1}(i) \end{bmatrix} \hat{P}_c^{-1}(i). \qquad (3.87)$$

By the process of pole assignment[27], we know that

$$\hat{P}_c^{-1}(i)\tilde{A}(i)\hat{P}_c(i) = \hat{P}_c^{-1}(i)A(i)\hat{P}_c(i) - \hat{P}_c^{-1}(i)b(i)K(i)\hat{P}_c(i)$$
$$= \hat{C}_c(\lambda, \lambda_1).$$

That is,

$$\tilde{A}(i) = \hat{P}_c(i)\hat{C}_c(\lambda, \lambda_1)\hat{P}_c^{-1}(i).$$

By Lemma 3.6.1, we have

$$\tilde{A}(i) = \hat{P}_c(i)\hat{Q}_c(\lambda, \lambda_1)\hat{D}_c(\lambda, \lambda_1)\hat{Q}_c^{-1}(\lambda, \lambda_1)\hat{P}_c^{-1}(i).$$

$\square$

Define

$$\Lambda(\lambda_1, t) = \begin{bmatrix} e^{\lambda_1 t} & te^{\lambda_1 t} & \cdots & t^{r-1}/(r-1)!e^{\lambda_1 t} \\ 0 & e^{\lambda_1 t} & \cdots & t^{r-2}/(r-2)!e^{\lambda_1 t} \\ \vdots & \vdots & \ddots & \vdots \\ 0 & 0 & \cdots & e^{\lambda_1 t} \end{bmatrix}, \qquad (3.88)$$

$$\Xi(i, \lambda, \lambda_1, t) = \hat{P}_c(i)\hat{Q}_c(\lambda, \lambda_1)\Lambda(\lambda_1, t)\hat{Q}_c^{-1}(\lambda, \lambda_1)\hat{P}_c^{-1}(i). \qquad (3.89)$$

We shall now derive some other supporting results.

Principle to select λ_1:   *For any $\beta_1 > 1$, $t_0 > 0$ and λ, select λ_1 with sufficiently negative real part, such that for all $i = 1, 2, \cdots, n$ and for all $t \geq t_0$,*

$$\lambda_{max}(\Xi^T(i, \lambda, \lambda_1, t)\Xi(i, \lambda, \lambda_1, t)) < \frac{1}{\beta_1}. \tag{3.90}$$

β_1 can be chosen to give some flexibility in determining the switching time of the controller.

Then, a key problem is whether there exists a λ_1 satisfying the above requirement. The following three lemmas are used to solve this problem.

Lemma 3.6.3. *For any desired decay rate $\lambda, i \in \{1, 2, \cdots, n\}$ and $t > 0$, there exists $\tilde{\lambda}_2(i, \lambda, t) < 0$ such that*

$$\lambda_{max}(\Xi^T(i, \lambda, \tilde{\lambda}_2(i, \lambda, t), t)\Xi(i, \lambda, \tilde{\lambda}_2(i, \lambda, t), t)) < \frac{1}{\beta_1}.$$

Proof. For any given $\lambda, i \in \{1, 2, \cdots, n\}$, and $t > 0$, it can be shown that

$$\lim_{\lambda_1 \to -\infty} \lambda_{max}(\Xi^T(i, \lambda, \lambda_1, t)\Xi(i, \lambda, \lambda_1, t)) = 0.$$

Thus, there exists a $\lambda_1(i, \lambda, t)$ that is dependent on i, λ and t, such that when $\lambda_1 < \lambda_1(i, \lambda, t)$,

$$\lambda_{max}(\Xi^T(i, \lambda, \lambda_1, t)\Xi(i, \lambda, \lambda_1, t)) < \frac{1}{\beta_1}.$$

Let

$$\tilde{\lambda}_2(i, \lambda, t) = \lambda_1(i, \lambda, t) - 1.$$

It follows that

$$\lambda_{max}(\Xi^T(i, \lambda, \tilde{\lambda}_2(i, \lambda, t), t)\Xi(i, \lambda, \tilde{\lambda}_2(i, \lambda, t), t)) < \frac{1}{\beta_1}.$$

$\square$

If the duration time of CVS i is known as $\Delta\tau_i$, the method of selecting λ_1 is given in Lemma 3.6.3. If we only know the upper bound of the duration time, the problem will be solved by Lemma 3.6.4; otherwise, it can be solved by Lemma 3.6.5.

Lemma 3.6.4. *Given any λ, there exists a $\hat{\lambda}_2 < 0$ such that when $\lambda_1 < \hat{\lambda}_2$, the inequality*

$$\lambda_{max}(\Xi^T(i, \lambda, \lambda_1, t)\Xi(i, \lambda, \lambda_1, t)) < \frac{1}{\beta_1} \tag{3.91}$$

holds for $t \in [(1 - \beta_2)\Delta_{1,i}, \Delta_{2,i}], i \in \{1, 2, \cdots, n\}$, where $\Delta_{2,i}$ is the upper bound of the time interval for switching and β_2 is a positive constant less than 1.

Proof. For any λ and $t_0 > 0$, by Lemma 3.6.3, there exists a $\lambda_1(i, \lambda, t_0)$ such that when $\lambda_1 < \lambda_1(i, \lambda, t_0)$,

$$\lambda_{max}(\Xi^T(i, \lambda, \lambda_1, t_0)\Xi(i, \lambda, \lambda_1, t_0)) < \frac{1}{\beta_1}.$$

It follows that

$$\lambda_{max}(\Xi^T(i, \lambda, \lambda_1(i, \lambda, t_0) - 1, t_0)\Xi(i, \lambda, \lambda_1(i, \lambda, t_0) - 1, t_0)) < \frac{1}{\beta_1}.$$

By the continuity of the eigenvalues of $\Xi^T(i, \lambda, \lambda_1, t)\Xi(i, \lambda, \lambda_1, t)$ with respect to t, we can find an open neighborhood $W(t_0)$ of t_0 satisfying the following condition:

$$\lambda_{max}(\Xi^T(i, \lambda, \lambda_1(i, \lambda, t_0) - 1, t)\Xi(i, \lambda, \lambda_1(i, \lambda, t_0) - 1, t)) < \frac{1}{\beta_1}.$$

i.e. for each $t \in W(t_0)$.

Consequently, for each $t \in [(1 - \beta_2)\Delta_{1,i}, \Delta_{2,i}]$, we generate an open cover of $[(1 - \beta_2)\Delta_{1,i}, \Delta_{2,i}]$ by taking the union of the set $W(t_0)$ as t_0 ranges over $[(1 - \beta_2)\Delta_{1,i}, \Delta_{2,i}]$. Using the compactness of $[(1 - \beta_2)\Delta_{1,i}, \Delta_{2,i}]$, we can extract a finite set of $\lambda_1(i, \lambda, t_j)_{j=1}^{N_1}$, where N_1 is the number of the finite open cover of $[(1 - \beta_2)\Delta_{1,i}, \Delta_{2,i}]$.

To complete the construction of $\hat{\lambda}_2$, we simply take $\hat{\lambda}_2$ to be the minimum one as given below

$$\hat{\lambda}_2 = \min_{1 \leq i \leq n} \min_{1 \leq j \leq N_1} \{\lambda_1(i, \lambda, t_j) - 1\}.$$

$\square$

Lemma 3.6.5. *Given any desired decay rate* λ, *there exists a* $\tilde{\lambda}_2 < 0$ *such that when* $\lambda_1 < \tilde{\lambda}_2$, *the inequality*

$$\lambda_{max}(\Xi^T(i, \lambda, \lambda_1, t)\Xi(i, \lambda, \lambda_1, t)) < \frac{1}{\beta_1}. \tag{3.92}$$

holds for $t \geq (1 - \beta_2)\Delta_{1,i}$, $i \in \{1, 2, \cdots, n\}$.

Proof. Note that the equation

$$\lim_{(t, -\lambda_1) \to (\infty, \infty)} \lambda_{max}(\Xi^T(i, \lambda, \lambda_1, t)\Xi(i, \lambda, \lambda_1, t)) = 0$$

holds for any given desired decay rate λ.

It follows that there exists $M(\lambda) > 0$, such that when $t > M(\lambda)$ and $\lambda_1 < -M(\lambda)$,

$$\lambda_{max}(\Xi^T(i, \lambda, \lambda_1, t)\Xi(i, \lambda, \lambda_1, t)) < \frac{1}{\beta_1}.$$

By Lemma 3.6.4, there exists a $\hat{\lambda}_2$ such that when $\lambda_1 < \hat{\lambda}_2$, we have

$$\lambda_{max}(\Xi^T(i,\lambda,\lambda_1,t)\Xi(i,\lambda,\lambda_1,t)) < \frac{1}{\beta_1}$$

for all $(1-\beta_2)\Delta_{1,i} \leq t \leq M(\lambda)$.

Let

$$\tilde{\lambda}_2 = \min\{\hat{\lambda}_2, -M(\lambda) - 1\}.$$

It follows that when $\lambda_1 < \tilde{\lambda}_2$,

$$\lambda_{max}(\Xi^T(i,\lambda,\lambda_1,t)\Xi(i,\lambda,\lambda_1,t)) < \frac{1}{\beta_1}.$$

$\square$

Remark 3.6.1. By the proofs of Lemmas 3.6.3–3.6.5, we know that the method proposed in these three lemmas can be used for both finite and infinite switched linear systems. $\square$

Based on Lemmas 3.6.1–3.6.5, a full state feedback sub-controller is designed for each CVS as follows:

1. For a given desired decay rate λ, determine λ_1 based on Lemmas 3.6.3–3.6.5.

2. Compute $d_l(0 \leq l \leq r-1)$ and $a_l(i)(0 \leq l \leq r-1)$ via (3.83) and (3.85).

3. Compute $K(i)$ by using (3.87).

Since the continuous state is always available, the switching instance of the system can be known exactly. As mentioned earlier, a small delay is imposed on the switchings of the controller to identify the subsequent CVSs. Meanwhile, to constrain and reduce the overshoot caused by the delay in activating the right sub-controller for the active CVS, the switching time t_j^c for the sub-controllers is defined as

$$\begin{cases} t_j^c \in \Omega_c(j) & \text{if } \Omega_c(j) \neq \emptyset \\ t_j^c = t_j^s + \frac{\beta_2 \Delta_{1,i}}{2} & \text{if } \Omega_c(j) = \emptyset \end{cases} \tag{3.93}$$

where t_j^s is the jth switching instant of the system and

$$\Omega_c(j) = \{t \mid t \geq t_{j-1}^c + (1-\beta_2)\Delta_{1,i} \; ; \; |t - t_j^s| \leq \frac{\beta_2 \min_i\{\Delta_{1,i}\}}{2}$$
$$e^{2\lambda(t-t_0)}\|X(t_0)\|^2 \leq \|X(t)\|^2 \leq \beta_1 e^{2\lambda(t-t_0)}\|X(t_0)\|^2\} \tag{3.94}$$

with $t_0^s = t_0$, where t_j^c is the jth switching instance of the controller.

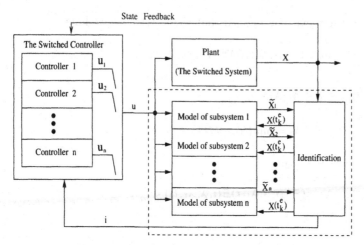

Fig. 3.9. The identification of the active CVS

Remark 3.6.2. In the set (3.94), the key idea is to restrict $\|X(t_j^c)\|^2 \leq \beta_1 e^{2\lambda(t_j^c - t_0)}\|X(t_0)\|^2$ when $t_j^c > t_j^s$. □

The initial CVS and its subsequent CVS of the system can be determined by using the state knowledge of the system and each CVS within the interval $[t_j^s, t_j^c]$. Obviously, there is only one CVS whose state can match the system exactly within this interval.

The whole process is illustrated as in Fig. 3.9, and the design procedures are summarized as follows:

1. Design a sub-controller for each CVS according to Lemmas 3.6.1–3.6.5.

2. Identify the initial CVS of the system by using the state knowledge of system (3.80) and the models of the CVSs within $[t_0, t_0^c]$ and switch the sub-controller from $u(t) = 0$ to the sub-controller corresponding to the initial active subsystem.

3. The CVS within $[t_j^s, t_{j+1}^s]$ is determined by using the state knowledge of system (3.80) and the models of the CVSs within $[t_j^s, t_j^c]$. The sub-controller corresponding to the active subsystem is applied within $[t_j^c, t_{j+1}^c]$.

Then we have the following result:

Theorem 3.6.1. *Consider the linear switched systems (3.80) satisfying Assumption 3.3, for any desired λ, there exists $M > 0$ and linear controller $u(t) = K(i)X(t)$ with the switching time defined in (3.93) and (3.94) such that for any $X(t_0)$ and all $t \in [t_0, \infty)$,*

$$\|X(t)\|^2 \leq M e^{2\lambda(t-t_0)}\|X(t_0)\|^2. \tag{3.95}$$

Proof. Assume that the switching instances of the system are $t_1^s, t_2^s, \cdots, t_g^s$ and the switching instances of the controller are $t_1^c, t_2^c, \cdots, t_g^c$, and $t > t_g^c$. For any λ, if the bound of the time interval between two switchings is known as $\Delta_{2,i}$, a λ_1 is selected in the way given by Lemma 3.6.3, otherwise, a λ_1 is chosen in the way given by Lemma 3.6.5. For the CVSs i, λ and λ_1, $K(i)$ is constructed as in Lemma 3.6.2.

Assume that the CVS of the system is l_j within the interval $[t_{j-1}^s, t_j^s]$. The results that are given in a) and b) below can be easily shown:

a) The inequality

$$t_j^c > t_j^s \tag{3.96}$$

holds for any j.

b) For any j, two inequalities

$$\|X(t_j^c)\|^2 \leq \beta_1 e^{2\lambda(t_j^c - t_0)} \|X(t_0)\|^2, \tag{3.97}$$

$$\|X(t_j^s)\|^2 \leq e^{2\lambda(t_j^s - t_0)} \|X(t_0)\|^2, \tag{3.98}$$

hold for both $\Omega_c(j) = \emptyset$ and $\Omega_c(j) \neq \emptyset$.

Now, it can be shown from Lemmas 3.6.3– 3.6.5 and the results in a) and b) that

$$\|X(t)\|^2 \leq M_1 e^{2\lambda(t - t_g^c)} \|X(t_g^c)\|^2, \tag{3.99}$$

where if the bound of the interval is known as $\Delta_{2,i}$, M_1 is given by

$$M_1 = \max_{1 \leq i \leq n} \max_{0 \leq t \leq \Delta\tau_i} \lambda_{max}(\Xi^T(i, \lambda, \lambda_1, t)\Xi(i, \lambda, \lambda_1, t)).$$

Otherwise it is given by

$$M_1 = \max_{1 \leq i \leq n} \max_{0 \leq t \leq \infty} \lambda_{max}(\Xi^T(i, \lambda, \lambda_1, t)\Xi(i, \lambda, \lambda_1, t)).$$

By Lemma 3.6.5, we know that

$$\max_{1 \leq i \leq n} \max_{0 \leq t \leq \infty} \lambda_{max}(\Xi^T(i, \lambda, \lambda_1, t)\Xi(i, \lambda, \lambda_1, t))$$

is bounded for the given decay λ and λ_1.

Let

$$M_2 = \max_{0 < t \leq \beta_2 \min_i\{\Delta_{1,i}\}} \max_{i \neq j} \|e^{(A(i) + b(i)K(j))t}\|^2 e^{-2\lambda\beta_2\Delta_{1,i}}.$$

Note that

$$t_{g+1}^c - t_{g+1}^s \leq \beta_2 \Delta_{1,i}.$$

It follows that for any $t_{g+1}^c > t > t_{g+1}^s$,

$$\|X(t)\|^2 \le M_2 e^{2\lambda(t-t_{g+1}^s)}\|X(t_{g+1}^s)\|^2. \qquad (3.100)$$

Using inequalities (3.93), (3.94), (3.97), and (3.99), it can be shown that when $t_{g+1}^s > t > t_g^c$,

$$\|X(t)\|^2 \le M_1 e^{2\lambda(t-t_g^c)}\|X(t_g^c)\|^2$$
$$< \beta_1 M_1 e^{2\lambda(t-t_0)}\|X(t_0)\|^2.$$

It can also be derived from inequalities (3.98) and (3.100) that when $t_{g+1}^c > t > t_{g+1}^s$,

$$\|X(t)\|^2 \le M_2 e^{2\lambda(t-t_{g+1}^s)}\|X(t_{g+1}^s)\|^2$$
$$< M_2 e^{2\lambda(t-t_0)}\|X(t_0)\|^2.$$

Thus

$$\|X(t)\|^2 \le \max\{M_2, \beta_1 M_1\}e^{2\lambda(t-t_0)}\|X(t_0)\|^2.$$

$\square$

Example 3.6.1. Consider a linear switched system which is composed of

$$\text{CVS 1};\quad A(1) = \begin{bmatrix} 1 & 0 \\ 1 & 1 \end{bmatrix}; \ b(1) = \begin{bmatrix} 1 \\ 0 \end{bmatrix},$$

$$\text{CVS 2};\quad A(2) = \begin{bmatrix} 1 & 1 \\ 0 & 1 \end{bmatrix}; \ b(2) = \begin{bmatrix} 0 \\ 1 \end{bmatrix}.$$

The initial state of the system is $X(0) = [100\ 100]^T$ and the initial CVS for the system is CVS 1.

1. Consider the case where the duration times of CVSs 1 and 2 are 3 and 4 seconds, respectively.

   Given $\lambda = -1$, select $\lambda_1 = -1$ according to Lemma 3.6.3. Then it can be shown that

   $$\lambda_{max}(\Xi^T(1,-1,-1,3)\Xi(1,-1,-1,3)) < 1,$$
   $$\lambda_{max}(\Xi^T(2,-1,-1,4)\Xi(2,-1,-1,4)) < 1.$$

   Thus, by Theorem 3.6.1,

   $$\|X(t)\|^2 < M e^{-2t}\|X(0)\|^2.$$

   Using equations (3.83)-(3.87), the switched controllers are computed as

   $$k(1) = \begin{bmatrix} 6 & 9 \end{bmatrix}, \text{ and } k(2) = \begin{bmatrix} 9 & 6 \end{bmatrix}.$$

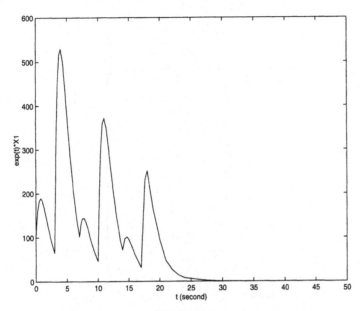

Fig. 3.10. The responses $e^t X_1(t)$ of the system

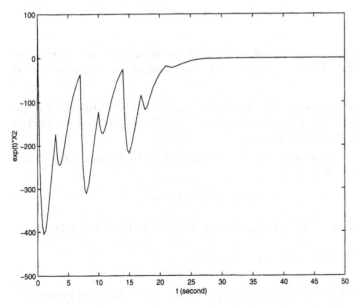

Fig. 3.11. The responses $e^t X_2(t)$ of the system

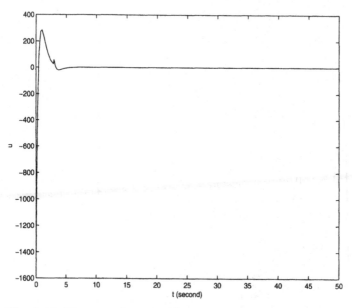

Fig. 3.12. The control input $u(t)$ of the system

The time responses and control input of the linear switched system are given in Figs. 3.10-3.12. Clearly, the states converges exponentially to zero.

2. Consider the case where the system switches from CVS 1 to CVS 2 if the state of the system satisfies that $|X_1(t)| = 0.5 * |X_2(t)|$, and it switches from CVS 2 to CVS 1 when $|X_1(t)| = 2 * |X_2(t)|$.

   Given $\lambda = -1$, we select $\lambda_1 = -0.5$ according to Lemma 3.6.5. Then from Theorem 3.6.1,

$$\|X(t)\|^2 < Me^{-2t}\|X(0)\|^2.$$

   Using equations (3.83)-(3.87), the switched controllers are computed as

$$k(1) = \begin{bmatrix} 5 & 6.25 \end{bmatrix}, \text{ and } k(2) = \begin{bmatrix} 6.25 & 5 \end{bmatrix}.$$

   The time responses and control input of the linear switched system are presented in Figs 3.13-3.15. Obviously, the states converges exponentially to zero.

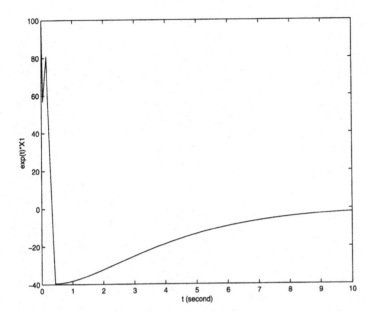

Fig. 3.13. The responses $e^t X_1(t)$ of the system

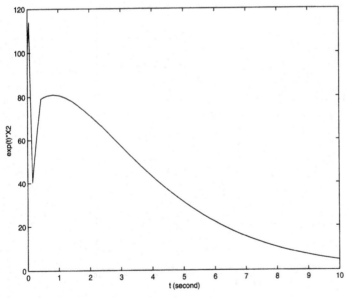

Fig. 3.14. The responses $e^t X_2(t)$ of the system

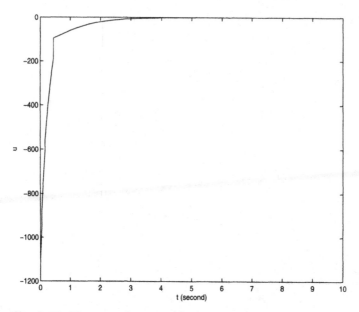

Fig. 3.15. The control input $u(t)$ of the system

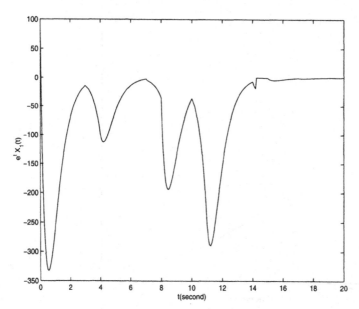

Fig. 3.16. The responses $e^t X_1 t)$ of the system using the second switching law with $\beta_1 = 10$

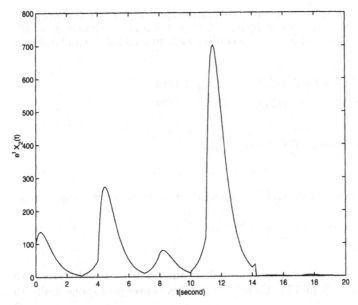

Fig. 3.17. The responses $e^t X_2 t)$ of the system with the second switching law with $\beta_1 = 10$

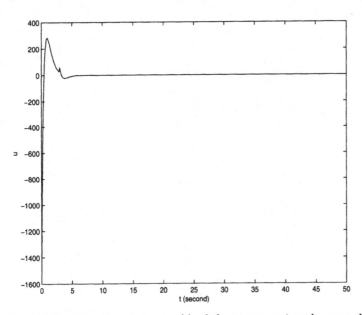

Fig. 3.18. The control input $u(t)$ of the system using the second switching law with $\beta_1 = 10$

3. Consider again the case where the duration times of CVSs 1 and 2 remain at 3 and 4 seconds, respectively. Let $\beta_2 = 0.4$ and we choose $\beta_1 = 10$. Given $\lambda = -1$, select $\lambda_1 = -2$ according to Lemma 3.6.3, it can be shown that

$$\lambda_{max}(\Xi^T(1, -1, -2, 3)\Xi(1, -1, -2, 3)) < 1/10,$$
$$\lambda_{max}(\Xi^T(2, -1, -2, 4)\Xi(2, -1, -2, 4)) < 1/10.$$

Thus, by Theorem 3.6.1, we know that

$$\|X(t)\|^2 < Me^{-2t}\|X(0)\|^2.$$

Using equations (3.83)-(3.87), the switched controllers are computed as

$$k(1) = \begin{bmatrix} 8 & 16 \end{bmatrix}, \text{ and } k(2) = \begin{bmatrix} 16 & 8 \end{bmatrix}.$$

The time responses and control input of the linear switched system are given in Figs. 3.16-3.18. Clearly, the states converges exponentially to zero. □

4. Stability of Nonlinear Switched and Impulsive Systems

In this chapter, we shall study the stability of nonlinear switched and impulsive systems of the form

$$\begin{cases} \dot{X}(t) = f(X(t), m(t)) \ ; \ t \neq t_j \\ X(t_j^+) = h(X(t_j^-), m(t_j^-), m(t_j^+)) \ . \\ m(t_j^+) \in \psi(m(t_i^-)) \end{cases} \qquad (4.1)$$

Recall that there are n CVSs, i.e. $m(t) \in \{1, 2, \cdots, n\}$. When $m(t) = i$, we simply denote the system functions as $f(X(t), i)$, $h(X(t), i, j)$, $\psi(i)$ etc.

We shall present three methods, namely multiple Lyapunov functions, linear approximation and generalized matrix measures for the stability analysis of the above class of nonlinear systems.

4.1 Multiple Lyapunov Functions

In this section, multiple Lyapunov functions are used to derive some sufficient conditions for the stability of nonlinear switched and impulsive systems.

Theorem 4.1.1. *A nonlinear switched and impulsive system of the form (4.1) is uniformly stable in the sense of Lyapunov with respect to an invariant set $\bar{S}_M$ if there exist Lyapunov functions $V(X(t), m(t))$, such that the following conditions hold.*

Condition 1 *There exist two positive numbers c_1 and c_2 such that the Lyapunov functions are bounded by*

$$c_1 \rho^2(S(t), \bar{S}_M) \leq V(X(t), m(t)) \leq c_2 \rho^2(S(t), \bar{S}_M). \qquad (4.2)$$

Condition 2 *There exists a number $c_3(i)$ such that*

$$\frac{\partial V(X(t), i)}{\partial X(t)} f(X(t), i) \leq c_3(i) \rho^2(S(t), \bar{S}_M). \qquad (4.3)$$

Condition 3 *For each pair of $m(t_j^-)$ and $m(t_j^+)$, there exists a positive number $\varepsilon_{m(t_j^-),m(t_j^+)}$ such that*

$$V(h(X(t_j^-)), m(t_j^-), m(t_j^+)), m(t_j^+))$$
$$\leq \varepsilon_{m(t_j^-),m(t_j^+)} V(X(t_j^-), m(t_j^-)). \tag{4.4}$$

Condition 4 *Lyapunov functions are non-increasing along cycle $LC(j)$ $(j = 1, 2, \cdots, \theta_0)$ of the form $v_1, v_2, \cdots, v_{k_0(j)}, v_1$, i.e,*

$$G(j) \triangleq \frac{\ln(\prod_{1 \leq l \leq k_0(j)} \frac{c_2}{c_1} \varepsilon_{v_l,v_{l+1}} e^{c_3(v_l)\bar{\Delta}_{v_l}})}{2} \leq 0, \tag{4.5}$$

where

$$\bar{\Delta}_i = \begin{cases} \Delta_{2,i}/c_1; \ i \in \Gamma_1 \\ \Delta_{1,i}/c_2; \ i \in \Gamma_2 \end{cases},$$
$$\Gamma_1 = \bar{M} - \Gamma_2,$$
$$\Gamma_2 = \{i | c_3(i) \leq 0\},$$

and $\Delta_{1,i}$ and $\Delta_{2,i}$ are as defined in (2.7) and (2.8).

Proof. Using Lemmas 2.2.1 and 2.2.2, it can be easily shown that

1. For any i, there exists an $\omega_i \in C[R^+, r^+]$ with $\omega_i(0) = 0$ such that

$$\sup_{t_{s,i}^k < t \leq t_{f,i}^k} V(X(t), m(t)) \leq \omega_i(V(X(t_{s,i}^k), m(t_{s,i}^k))),$$

where

$$\omega_i(z) = \begin{cases} e^{\frac{c_3(i)\Delta_{2,i}}{c_1}} z; \ i \in \Gamma_1 \\ z; \qquad i \in \Gamma_2 \end{cases}. \tag{4.6}$$

2. For any t_k^+ of $\Sigma|_t$ (2.6), we have

$$V(X(t_k^+), m(t_k^+)) \leq \hbar_1^{2n-2} \hbar_2^2 V(X(t_0), m(t_0)),$$

where

$$\hbar_1 = \max\{1, \max_{i,l}\{\sqrt{\varepsilon_{i,l}} \frac{c_2}{c_1}\}\}, \tag{4.7}$$

$$\hbar_2 = \prod_{i \in \Gamma_1} [e^{\frac{c_3(i)\Delta_{2,i}}{c_1}}]. \tag{4.8}$$

In other words, $\nu(z) = \hbar_1^{2n-2} \hbar_2^2 z$.

By Theorem 2.5.1, the nonlinear switched and impulsive system is uniformly stable in the sense of Lyapunov with respect to an invariant set $\bar{S}_M$.   □

Remark 4.1.1. Condition (4.3) implies that the state of each unstable CVS is finite within a finite interval and condition (4.5) means that Lyapunov functions are non-increasing along each type of cycle.   □

Remark 4.1.2. By analyzing the discrete behavior of the nonlinear switched and impulsive systems, we are able to identify the finite types of cycles and study the continuous behavior of the systems along each type of cycle. In this way, the non–increasing subsequence of the "switching" $t_{s,i_0}^1, t_{s,i_0}^2, \cdots, t_{s,i_0}^k,$ $\cdots$ can be identified and the continuous functions $\omega_i(z)$, $i = 1, 2, \cdots, n$ can also be constructed to bound the Lyapunov functions.   □

Remark 4.1.3. Note that $\varepsilon_{m(t_j^-), m(t_j^+)}$ is not required to be less than or equal to 1. Thus, our results are less conservative than that of [10], which requires the Lyapunov function to be non-increasing along each "switching" instance.   □

Remark 4.1.4. Note that the Lyapunov function is not required to be non–increasing along each CVS and the nonlinear switched and impulsive system is not required to be periodic. Thus, the results presented here are less conservative than that of [96].   □

Theorem 4.1.2. *A nonlinear switched and impulsive system of the form (4.1) is uniformly asymptotically stable in the sense of Lyapunov with respect to an invariant set $\bar{S}_M$ if there exist Lyapunov functions $V(X(t), m(t))$, such that the Conditions 1–3 of Theorem 4.1.1 hold, and the following condition holds.*

Condition 4' *Lyapunov functions are non-increasing along cycle $LC(j)$ $(j = 1, 2, \cdots, \theta_0)$, i.e.*

$$G(j) \triangleq \frac{\ln(\prod_{1 \leq l \leq k_0(j)} \frac{c_2}{c_1} \varepsilon_{v_l, v_{l+1}} e^{c_3(v_l)\bar{\Delta}_{v_l}})}{2} < 0. \tag{4.9}$$

Proof. Using Lemmas 2.2.1 and 2.2.2 and Condition 4', it can be easily shown that

1. There exists an $\omega_i \in C[R^+, R^+]$ with $\omega_i(0) = 0$ such that

$$\sup_{t_{0,i}^l < t \leq t_{f,i}^l} V(X(t), m(t)) \leq \omega_i(V(X(t_{0,i}^l), m(t_{0,i}^l))),$$

where $\omega_i(z)$ is given in equation (4.6).

2. There exists a subsequence $t^1_{s,i_0}, t^2_{s,i_0}, \cdots, t^k_{s,i_0}, \cdots$ of $\Sigma|_t$ (2.6) which satisfies that

$$V(X(t^{k+1}_{s,i_0}), m(t^{k+1}_{s,i_0})) \leq \mathrm{L}^{2\aleph_{t^k_{s,i_0}, t^{k+1}_{s,i_0}}} V(X(t^k_{s,i_0}), m(t^k_{s,i_0})) ; \; k = 1, \cdots \quad (4.10)$$

where $\aleph_{t^k_{s,i_0}, t^{k+1}_{s,i_0}}$ is the total number of time that each cycle appears in the path from $m(t^k_{s,i_0})$ to $m(t^{k+1}_{s,i_0})$, and

$$\mathrm{L} = \max_{1 \leq j \leq \theta_0} \{e^{G(j)}\}. \quad (4.11)$$

3. There exists a $\Phi \in C[R^+, R^+]$ with $\Phi(0) = 0$ such that for each $t^{k+1}_{s,i_0} > t^+_l > t^k_{s,i_0}$,

$$V(X(t^+_l), m(t^+_l)) \leq \Phi(V(X(t^k_{s,i_0}), m(t^k_{s,i_0}))),$$

where

$$\Phi(z) = \hbar^{2n-2}_1 \hbar^2_2 \mathrm{L}^{2\aleph_{t^k_{s,i_0}, t^+_l}} z.$$

4. There exists a $\Psi \in C[R^+, R^+]$ with $\Psi(0) = 0$ such that

$$V(X(t^+_l), m(t^+_l)) \leq \Psi(X(t_0), m(t_0)) ; \; t_0 \leq t^+_l \leq t^1_{s,i_0},$$

where

$$\Psi(z) = \hbar^{2n-2}_1 \hbar^2_2 \mathrm{L}^{2\aleph_{t_0, t_l^+}} z.$$

By Theorem 2.5.2 , we know that a nonlinear switched and impulsive system of the form (4.1) is uniformly asymptotically stable in the sense of Lyapunov with respect to an invariant set $\bar{S}_M$. □

Remark 4.1.5. Hou et al. [63] derived some sufficient conditions for such a nonlinear switched and impulsive system to be uniformly asymptotically stable with respect to an equilibrium point. However, their stability results are only local and the knowledge of the continuous trajectory is required. □

Remark 4.1.6. Dayawansa and Martin [44] obtained some conditions for the uniformly asymptotic stability of nonlinear switched and impulsive systems in the case where the switchings of the CVSs are arbitrary. Every CVS is required to be stable and Lyapunov functions should be non-increasing along the whole switching sequence. The results presented do not require every CVS to be stable because switchings are governed by a finite automata. Moreover, the Lyapunov functions are only required to be non-increasing along each type of cycle. □

Remark 4.1.7. Note that He and Lemmon [91] also tried to use the "cycle analysis method" to derive some less conservative conditions for the Lyapunov stability of nonlinear switched and impulsive systems. However, their emphasis is on the discrete part of the system. The Lyapunov functions are still required to be always non-increasing along each CVS in their results. Thus, their results can only be used to study the stability of nonlinear switched and impulsive systems that are totally composed of stable CVSs. □

A numerical example is now used to illustrate the application of the multiple Lyapunov functions method.

Example 4.1.1. Consider a nonlinear switched and impulsive system which is composed of three CVSs. The switchings of the CVSs are illustrated in Fig. 3.6. The other specifications are as follows:

1. The reset maps are

$$h(\cdot, 1, 2) = \begin{cases} 2X_1(t) + X_2(t)/2 \\ X_1(t)/2 + 2X_2(t) \end{cases},$$

$$h(\cdot, 2, 1) = \begin{cases} X_1(t) \\ X_2(t) \end{cases},$$

$$h(\cdot, 1, 3) = \begin{cases} X_1(t) \\ X_2(t) \end{cases},$$

$$h(\cdot, 3, 1) = \begin{cases} X_1(t)/2 + X_2(t)2/3 \\ X_1(t)/2 + X_2(t)1/3 \end{cases}.$$

2. The switching conditional sets are

$$S_{1,2} = S_{1,3} = \{t_{f,1}^k - t_{s,1}^k = 2 ; \forall k\} ;$$
$$S_{2,1} = \{t_{f,2}^k - t_{s,2}^k = 3 ; \forall k\} ;$$
$$S_{3,1} = \{t_{f,3}^k - t_{s,3}^k = 4 ; \forall k\}.$$

3. The state equations of the CVSs are

$$f(X(t), 1) = \begin{cases} (1/8 + \sin^2(X_1(t))/2 + \sin(X_1(t))/2)X_1(t) - X_2(t) \\ X_1(t) + (1/8 + \cos^2(X_1(t))/2 + \cos(X_1(t))/2)X_2(t) \end{cases},$$

$$f(X(t), 2) = \begin{cases} (-3 + \sin^2(X_1(t))X_1(t) \\ (-4 + \sin^2(t))X_2(t) \end{cases},$$

$$f(X(t), 3) = \begin{cases} (-3 + |\cos(X_2(t))|)X_1(t) + 2X_2(t) \\ -2|\sin(X_1(t))|X_1(t) + (-4 + |\cos(X_2(t))|)X_2(t) \end{cases}.$$

Select the Lyapunov function as $V(X(t), m(t)) = X_1^2(t) + X_2^2(t)$. The conditions of Theorem 4.1.2 are verified as follows:

Condition 1: $c_1 = c_2 = 1$.

Condition 2: It can be easily verified that

$$1/4V(X(t),1) \le \frac{\partial V(X(t),1)}{\partial X(t)} f(X(t),1) \le 9/4V(X(t),1), \tag{4.12}$$

$$\frac{\partial V(X(t),2)}{\partial X(t)} f(X(t),2) \le -4V(X(t),2),$$

$$\frac{\partial V(X(t),3)}{\partial X(t)} f(X(t),3) \le -3V(X(t),3).$$

Condition 3: It can be shown that

$$9/4V(X(t),1) \le V(h(X(t),1,2),2) \le 25/4V(X(t),1), \tag{4.13}$$
$$V(X(t),1) \le V(h(X(t),1,3),3) \le 5V(X(t),1), \tag{4.14}$$
$$V(h(X(t),3,1),1) \le 2V(X(t),3),$$
$$V(h(X(t),2,1),1) \le 2V(X(t),2).$$

Condition 4'. For $LC(1) = (CVS1, CVS3, CVS1)$, we have

$$e^{-3 \times 4 + 9/4 \times 2} \times 5 \times 2 < 1/2 < 1,$$

and for $LC(2) = (CVS1, CVS2, CVS1)$, we obtain

$$e^{-4 \times 3 + 9/4 \times 2} \times \frac{25}{4} \times 2 < 1/2 < 1.$$

By Theorem 4.1.2, the nonlinear switched and impulsive system in this example is uniformly asymptotically stable with respect to $X_e = 0$.   □

Remark 4.1.8. From (4.12)–(4.14), we have

$$V(X(t_{2k+1}^+), m(t_{2k+1}^+)) \ge e^{1/2} V(X(t_{2k}^+), m(t_{2k}^+)) > V(X(t_{2k}^+), m(t_{2k}^+)) \quad k = 1, 2, \cdots$$

That is, the Lyapunov function is not non-increasing along the whole sequence of the "switchings". Moreover, CVS 1 is unstable. Thus, the results of [153, 21, 156] cannot be used to consider the stability of the nonlinear switched and impulsive system in this example.   □

4.2 Generalized Matrix Measure

Similar to the linear case, a simple tool of generalized matrix measure, is now introduced to the stability analysis of nonlinear switched and impulsive systems.

The system (4.1) can be regarded as an r-dimensional fluid flow, where $\dot{X}$ is the r-dimensional "velocity" vector at the r-dimensional position X. Then equations (4.1) yield the differentiable relations [117]:

$$\frac{d\delta X(t)}{dt} = \frac{\partial f}{\partial X}(X(t), m(t))\delta X(t), \tag{4.15}$$

and the difference relations

$$\delta X(t_j^+) = \frac{\partial h}{\partial X}(X(t_j^-), m(t_j^-), m(t_j^+))\delta X(t_j^-), \tag{4.16}$$

where δX is a virtual displacement which is an infinitesimal displacement at a fixed time. Note that the notion of virtual displacement, pervasive in physics and in the calculus of variation, is also a well defined mathematical object. Generally, δX and $\delta X^T \delta X$ are differentiable with respect to time [117].

In this section, sufficient conditions are derived on the stability of nonlinear switched and impulsive systems by considering the characteristics of $\delta X^T \delta X$ along each type of cycle.

It can be shown from equations (4.15) and (4.16) that

$$\frac{d\delta X^T \delta X}{dt} = \delta X^T(\frac{\partial f^T}{\partial X} + \frac{\partial f}{\partial X})\delta X,$$

and

$$\delta X^T(t_j^+)\delta X(t_j^+) = \delta X^T(t_j^-)\frac{\partial h^T}{\partial X}\frac{\partial h}{\partial X}\delta X(t_j^-). \tag{4.17}$$

The concept of generalized matrix measure is introduced to nonlinear systems as follows:

Definition 4.2.1. *For any continuous differentiable function $f(X)$, the generalized matrix measures of the function, $\mu(f, CS)$, $\mu(|f|_m, CS)$, in a compact set CS, are*

$$\mu(f, CS) = \sup_{X \in CS} \{\mu(\frac{\partial f}{\partial X})\}, \tag{4.18}$$

$$\mu(|f|_m, CS) = \sup_{X \in CS} \{\mu(|\frac{\partial f}{\partial X}|_m)\}. \tag{4.19}$$

The generalized matrix norms, $\|f\|_{CS}$, $\||f|_m\|_{CS}$, in a compact set CS, are

$$\|f\|_{CS} = \sup_{X \in CS} \{|\frac{\partial f}{\partial X}|\}, \tag{4.20}$$

$$\||f|_m\|_{CS} = \sup_{X \in CS} \{\||\frac{\partial f}{\partial X}|_m\|\}. \tag{4.21}$$

$\square$

Then, we have the following result:

Lemma 4.2.1. *For any continuously differentiable $h(x)$ and $f(x)$ satisfying $|\frac{\partial f}{\partial X}|_m \le A$ within a compact set CS and $a \ge 0$,*

$$\mu(f+g, CS) \le \mu(f, CS) + \mu(g, CS), \tag{4.22}$$

$$\mu(f, CS) \le \|f\|_{CS}, \tag{4.23}$$

$$\mu(af, CS) = a\mu(f, CS), \tag{4.24}$$

$$\mu(f, CS) \le \mu(|f|_m, CS) \le \mu(A), \tag{4.25}$$

$$\|f\|_{CS} \le \||f|_m\|_{CS} \le \|A\|. \tag{4.26}$$

Proof. We only prove inequality (4.22) as the other inequalities can be proven in a similar way.

For any given $X \in CS$,

$$\mu(\frac{\partial f}{\partial X}(X) + \frac{\partial g}{\partial X}(X)) \le \mu(\frac{\partial f}{\partial X}(X)) + \mu(\frac{\partial g}{\partial X}(X))$$
$$\le \mu(f, CS) + \mu(g, CS).$$

It follows that

$$\mu(f+g, CS) = \sup_{X \in CS} \{\mu(\frac{\partial(f+g)}{\partial X}(X))\}$$
$$= \sup_{X \in CS} \{\mu(\frac{\partial f}{\partial X}(X) + \frac{\partial g}{\partial X}(X))\}$$
$$\le \mu(f, CS) + \mu(g, CS).$$

$\square$

To study the stability of nonlinear switched and impulsive systems, we suppose that the following assumption holds.

Assumption 4.1 *There exists a compact set $\nabla \subseteq \bar{CS} = \cup_{i=1}^n \hat{CS}_i$ such that ∇ contains the given trajectory and for each $LC(j)(j = 1, 2, \cdots, \theta_0)$ of the form $v_1, v_2, \cdots, v_{k_0(j)}, v_1$,*

$$G(j) \triangleq \sum_{l=1}^{k_0(j)} \mu(f(X, v_l), \nabla)\Delta_{3,v_l} + \ln(\|h(X, v_l, v_{l+1})\|_{\hat{S}_{v_l, v_{l+1}}}) < 0, \tag{4.27}$$

where

$$\hat{CS}(i) = \{X | \frac{\partial f^T}{\partial X}(X, i) + \frac{\partial f}{\partial X}(X, i) \text{ is negative definite}\}, \tag{4.28}$$

$$\Delta_{3,v_l} = \begin{cases} \Delta_{1,v_l}; & v_l \in \Gamma_2 \\ \Delta_{2,v_l}; & v_l \in \Gamma_1 \end{cases}, \tag{4.29}$$

$$\Gamma_1 = \{i | \mu(f(X, i), \nabla) \ge 0\}; \ \Gamma_2 = \{i | \mu(f(X, i), \nabla) < 0\}. \tag{4.30}$$

$\square$

Remark 4.2.1. Assumption 4.1 implies that there exists a compact set such that $\delta X^T \delta X$ is non-increasing along each type of cycle when the state stays in the set. The compact set is the whole R^r space if each CVS is linear. Moreover, when $f(X, i) = A(i)X$, i.e. it is a linear function, $i \in \Gamma_1$ implies that the matrix measure of $A(i)$ is not less than 0 and $i \in \Gamma_2$ implies that the matrix measure of $A(i)$ is less than 0.   □

We now give a general method to check Assumption 4.1. Without loss of generality, we suppose that the given trajectory is $\tilde{X}(t)(t \geq t_0)$ and define

$$\hat{\alpha}_j(\tilde{X}(t)) = \sum_{i \in LC(j)} (\frac{\partial f}{\partial X}(\tilde{X}(t))\Delta_{3,i} + \|\frac{\partial h}{\partial X}\|_{\tilde{X}(t)}). \tag{4.31}$$

Check if there exists a ξ_0 such that the set $\{X(t) | \|X(t) - \tilde{X}(t)\| < \xi_0\}$ is a subset of $\bar{CS}$ and if $\hat{\alpha}_j(\tilde{X}(t)) < 0$ holds for all j. If so, then Assumption 4.1 holds. The reason for this is presented below.

Since $\hat{\alpha}_j(\tilde{X}(t)) < 0$, then there exists a $\xi_j (j = 1, 2, \cdots, \theta_0)$ such that when $\|X(t) - \tilde{X}(t)\| < \xi_j$, $\hat{\alpha}_j(X(t)) < 0$. Let $\xi = \frac{1}{2} \min_{0 \leq j \leq \theta_0}\{\xi_j\}$ and

$$SCS = \{X(t) | \|X(t) - \tilde{X}(t)\| \leq \xi\}. \tag{4.32}$$

Then Assumption 4.1 holds with SCS given in (4.32).

For a given system, there may be other better methods to find a larger compact set SCS to satisfy Assumption 4.1. As an example, an alternative method will be presented in a numerical example later on.

Proposition 4.2.1. *Assume that Assumption 4.1 holds. Consider a closed path CP_k with the starting time being t_{s,CP_k} and the ending time being t_{f,CP_k}. If $X(t) \in \nabla$ holds for all $t \in [t_{s,CP_k}, t_{f,CP_k}]$, then we have*

$$\|\delta X(t_{f,CP_k})\| \leq L^{\aleph_{t_{s,CP_k}, t_{f,CP_k}}}\|\delta X(t_{s,CP_k})\|, \tag{4.33}$$

and

$$\|\delta X(t_j^+)\| \leq \hbar_1^{n-1}\hbar_2\|\delta X(t_{s,CP_k})\|, \tag{4.34}$$

where $t_{s,CP_k} \leq t_j^+ \leq t_{f,CP_k}$, $\aleph_{t_{s,CP_k}, t_{f,CP_k}}$ is the total number of time that each cycle appears in the path from $m(t_{s,CP_k})$ to $m(t_{f,CP_k})$, and

$$L = \max_{1 \leq j \leq \theta_0} e^{G(j)}, \tag{4.35}$$

$$\hbar_1 = \max\{1, \max_{i,l}\{\|h(X, i, l)\|_{\hat{S}_{i,l}}\}\}, \tag{4.36}$$

$$\hbar_2 = \prod_{i \in \Gamma_1} e^{\mu(f(X,i), \nabla)\Delta_{2,i}}. \tag{4.37}$$

□

We shall now consider the stability of a nonlinear switched and impulsive system of the form (4.1).

Theorem 4.2.1. *A nonlinear switched and impulsive system of the form (4.1) is locally asymptotically stable with respect to a given trajectory if Assumption 4.1 holds.*

Proof. Suppose that the radius of the largest ball in ∇ is r. Let

$$r_0 = \frac{r}{\hbar_1^{n-1}\hbar_2 e^{\max_{i \in \Gamma_1}\{\mu(f(x,i),\nabla)\Delta_{2,i}\}}}.$$

Using Proposition 4.2.1, it can be easily proven that

1. For any t_j^-, if $X(t) \in \nabla$ holds for all $t_0 \leq t \leq t_j^-$, then

$$\|\delta X(t)\| \leq \hbar_1^{n-1}\hbar_2 L^{N_{t_0,t}}\|\delta X(t_0)\| \tag{4.38}$$

holds for any $t_0 \leq t \leq t_j^+$.

2. $X(t) \in \nabla$ for any t if $\|\delta X\| \leq r_0$ by induction.

From inequality (4.38) and Lemma 2.2.3, we have

$$\lim_{t \to \infty} \|\delta X(t)\| = 0.$$

That is, the result holds.   □

Consider the case that $h(X, i, l) = X$, that is, there is no impulsive switchings in the systems. In this case, Assumption 4.1 becomes

Assumption 4.2 *There exists a compact set $\nabla \subseteq \bar{CS}$ such that ∇ contains the given trajectory and*

$$G(j) = \sum_{l=1}^{k_0(j)} \mu(f(X, v_l), \nabla)\Delta_{3,v_l} < 0 \ ; \ j = 1, 2, \cdots, \theta_0. \tag{4.39}$$

□

From Theorem 4.2.1, we can obtain the following corollary:

Corollary 4.2.1. *A nonlinear switched system of the form (4.1) is locally asymptotically stable with respect to a given trajectory if Assumption 4.2 holds.*   □

We shall now present a numerical example to illustrate the application of generalized matrix measure.

Example 4.2.1. Consider the following nonlinear switched system that is composed of two CVSs:

$$f(X,1) = \begin{bmatrix} -2X_1 + 3X_1^2 \\ -3X_2 + X_2^2 \end{bmatrix},$$

$$f(X,2) = \begin{bmatrix} X_1 + X_1^2 \\ X_2/2 + 2X_2^3 \end{bmatrix}.$$

The dwell time of CVS 1 and CVS 2 are 6 and 2.5, respectively.

Note that

CVS 1: $\dfrac{d\delta X(t)}{dt} = \begin{bmatrix} -2 + 6X_1^2 & 0 \\ 0 & -3 + 2X_2 \end{bmatrix} \delta X(t),$

CVS 2: $\dfrac{d\delta X(t)}{dt} = \begin{bmatrix} 1 + 2X_1 & 0 \\ 0 & 1/2 + 6X_2^2 \end{bmatrix} \delta X(t).$

It follows that

$$\hat{C}S(1) = \{X|X_1 < 1/3 \,;\, X_2 < 3/2\},$$
$$\hat{C}S(2) = \emptyset.$$

Thus, $\bar{C}S = \hat{C}S(1)$. Let $\xi_0 = 1/6$. Obviously, the set $\{X|\|X\| < 1/6\}$ is a subset of $\bar{C}S$. It can also be checked that $\hat{a}(0) = -2 \times 6 + 2.5 \times 1 < 0$. Assumption 4.2 can be checked as in the following derivations.

Note that $\hat{a}(X) < 0$ when $\|X\| < \xi_1 \overset{\Delta}{=} 1/6$. Consider the following compact set:
$$\nabla = \{X|\|X\| \le 1/12\}.$$
From Definition 4.2.1, we know that

$$\bar{\mu}(f(\cdot,1),\nabla) = -1.5 \,;\, \bar{\mu}(f(\cdot,2),\nabla) = 1/6.$$

It can be shown that

$$6\bar{\mu}(f(\cdot,1),\nabla) + 2.5\bar{\mu}(f(\cdot,2),\nabla) < 0.$$

That is, Assumption 4.2 holds. From Corollary 4.2.1, we know that the nonlinear switched system is locally asymptotically stable with respect to $X_e = 0$.

It is also possible to find other type of compact sets to satisfy Assumption 4.2. Actually, consider the following compact set

$$\nabla = \{X|-1 \le X_1 \le 1/6 \,;\, -1/2 \le X_2 \le 1/2\}. \tag{4.40}$$

From Definition 4.2.1, we know that

$$\bar{\mu}(f(\cdot,1),\nabla) = -1 \,;\, \bar{\mu}(f(\cdot,2),\nabla) = 2.$$

It can be shown that

$$6\tilde{\mu}(f(\cdot,1),\nabla) + 2.5\tilde{\mu}(f(\cdot,2),\nabla) = -1 < 0.$$

That is, Assumption 4.2 holds.   □

The results obtained from these two sections show that Lyapunov function and generalized matrix measure convert a vector into a scalar. Since scalars are always commutative, sufficient conditions like (4.5), (4.27) and (4.39) can be derived to study the stability of nonlinear switched and impulsive systems. Moreover, these conditions can be easily checked.

4.3 Linear Approximation Method

In this section, linear approximation method is used to study the local stability of nonlinear switched and impulsive system (4.1). Assume that $f(0, m(t)) = 0$ and $h(0, m(t_j^-), m(t_j^+)) = 0$.

To find the linear approximation of the behavior of the system along each cycle, we need to consider the behavior of the switched and impulsive systems along each type of cycle $LC(j)(j = 1, 2, \cdots, \theta_0)$, $v_1, v_2, \cdots, v_{k_0(j)}, v_1$. From equation (4.1), we have

$$X(t) = e^{A(v_l)(t-t_{s,v_l}^k)}X((t_{s,v_l}^k) + \int_{t_{s,v_l}^k}^{t} e^{A(v_l)(t-\tau)}[f(X(t),v_l) - A(v_l)X(t)]d\tau$$

for all $t \in [t_{s,v_l}^k, t_{f,v_l}^k)$, $l = 1, 2, \cdots, n$, and

$$X(t_{s,v_{l+1}}^k) = D(v_l, v_{l+1})X(t_{f,v_l}^k) + [h(X(t_{f,v_l}^k), v_l, v_{l+1}) - D(v_l, v_{l+1})X(t_{f,v_l}^k)],$$

where $A(i)$ and $D(i,j)$ are the linear part of f and h, respectively, i.e.

$$A(i) = \frac{\partial f}{\partial X}(0, i) ; \ D(i,j) = \frac{\partial h}{\partial X}(0, i, j). \tag{4.41}$$

It follows that

$$X(t_{s,v_1}^{k+1}) = \Upsilon(j)X(t_{s,v_1}^k) + \sum_{w=1}^{k_0(j)} [\prod_{s=w+1}^{k_0(j)} (D(v_s, v_{s+1})e^{A(v_s)\Delta\tau_{v_s}})$$

$$(\int_{t_{s,v_w}^k}^{t_{s,v_{w+1}}^k} D(v_w, v_{w+1})e^{A(v_w)(t_{s,v_{w+1}}^k - \tau)}(f(X(t),v_w) - A(v_w)X(t))d\tau$$

$$+ h(X(t_{s,v_{w+1}}^k), v_w, v_{w+1}) - D(v_w, v_{w+1})X(t_{s,v_{w+1}}^k))]. \tag{4.42}$$

where $\Upsilon(j)$ is given in equation (3.35).

Then, the linear characteristics of the system along cycle $LC(j)$ can be represented by $\Upsilon(j)$. It was shown that the stability of nonlinear switched and impulsive systems is totally determined by the system behavior along each type of cycle. So, the stability of the original switched nonlinear system is also determined by $\Upsilon(j)$.

Suppose that

$$\|f(X,i) - A(i)X\| \le \min\{1,\xi\}\|X\| \; ; \; i = 1,2,\cdots,n \; ; \; \forall t,$$
$$\|h(X,i,l) - D(i,l)X\| \le \min\{1,\xi\}\|X\| \; ; \; i,l = 1,2,\cdots,n \; ; \; \forall t,$$

and all $G(j)(j = 1,2,\cdots,\theta_0)$ defined in equation (3.34) are less than 0, where ξ is an arbitrary positive number. Let $L = \max_{1 \le j \le \theta_0}\{e^{G(j)}\}$. Then, for any cycle $LC(j)$ of the form $\upsilon_1,\upsilon_2,\cdots,\upsilon_{k_0(j)},\upsilon_1$,

$$\|X(t_{s,\upsilon_1}^{k+1})\| \le \tilde{q}_0(\xi)\|X(t_{s,\upsilon_1}^k)\|,$$

where

$$\tilde{q}_0(\xi) = L + \prod_{i=1}^n (\hbar_1 e^{a_0 \Delta\tau_i} + \hbar_1 e^{(2a_0+1)\Delta\tau_i}\Delta\tau_i\xi + \xi e^{a_0\Delta\tau_i}$$
$$+ \xi^2 \Delta\tau_i e^{(2a_0+1)\Delta\tau_i})^n - \hbar_1^n e^{a_0 \sum_{i=1}^n \Delta\tau_i},$$
$$a_0 = \max_{1 \le i \le n}\{\mu(A(i))\},$$
$$\hbar_1 = \max\{1, \max_{1 \le i,l \le n}\{\|D(i,l)\|\}\}.$$

Note that $\tilde{q}_0(0) = L < 1$ and $\tilde{q}_0(\xi)$ is a continuous function of ξ, there exists a $\xi_0 > 0$ such that

$$\tilde{q}_0(\xi_0) < 1.$$

From (4.41), there exists a $\delta > 0$ such that when $\|X(t)\| < \delta$,

$$\|f(X,i) - A(i)X\| \le \min\{1,\xi_0\}\|X\| \; ; \; i = 1,2,\cdots,n, \forall t,$$
$$\|h(X,i,l) - D(i,l)X\| \le \min\{1,\xi_0\}\|X\| \; ; \; i,l = 1,2,\cdots,n \; ; \; \forall t.$$

We can then obtain the following propositions:

Proposition 4.3.1. *Consider cycle $LC(j)$ of the form $\upsilon_1,\upsilon_2,\cdots,\upsilon_{k_0(j)},\upsilon_1$. If $\|X(t_{s,\upsilon_1}^k)\| < \dfrac{e^{-(a_0+1)\sum_{i=1}^n \Delta\tau_i}\delta}{(1+\hbar_1)^n}$, then $\|X(t)\| < \delta$ holds for all $t \in [t_{s,\upsilon_1}^k, t_{s,\upsilon_1}^{k+1}]$.*

□

Proposition 4.3.2. *Consider cycle $LC(j)$ of the form $\upsilon_1,\upsilon_2,\cdots,\upsilon_{k_0(j)},\upsilon_1$. If $\|X(t)\| < \delta$ holds for all $t \in [t_{s,\upsilon_1}^k, t_{s,\upsilon_1}^{k+1}]$, then*

$$\|X(t_{s,\upsilon_1}^k)\| < \tilde{q}_0(\xi_0)\|X(t_{s,\upsilon_1}^{k+1})\|.$$

□

Proposition 4.3.3. *Consider any closed path with the first CVS and the last CVS as CVS i. If $\|X(t^k_{s,i})\| < \frac{e^{-2(a_0+1)\sum_{i=1}^{n} \Delta\tau_i}\delta}{(1+\hbar_1)^{2n}}$, then $\|X(t)\| < \delta$ holds for all $t \in [t^k_{s,i}, t^{k+1}_{s,i}]$.*   □

Proposition 4.3.4. *Consider any closed path with the first CVS and the last CVS as CVS i. If $\|X(t)\| < \delta$ holds for all $t \in [t^k_{s,i}, t^{k+1}_{s,i}]$, then*

$$\|X(t^k_{s,i})\| < \tilde{q}_0^{\aleph_{t^k_{s,i},t^{k+1}_{s,i}}}(\xi_0)\|X(t^{k+1}_{s,i})\|,$$

where $\aleph_{t^k_{s,i},t^{k+1}_{s,i}}$ is the total number of time that each cycle appears in the path from $m(t^k_{s,i})$ to $m(t^{k+1}_{s,i})$.   □

From the four propositions, we can get the following result:

Theorem 4.3.1. *A nonlinear switched system (4.1) with a fixed duration time for each CVS is locally asymptotically stable with respect to $X_e = 0$ if all $G(j)(j = 1, 2, \cdots, \theta_0)$ defined in equation (3.34) are less than 0.*

Proof. Define the Lyapunov function as

$$V(X(t), m(t)) = \|X(t)\|^2.$$

Since there are infinite numbers of switchings and the number of CVSs is n, there exists at least one CVS which appears infinite number of times. Assume that v_0 is the first such CVSs. Similar to the proof of Proposition 4.3.3, we know that when

$$\|X(0)\| < \frac{\frac{e^{-2(1+a_0)\sum_{i=1}^{n}\Delta\tau_i}\delta}{(1+\hbar_1)^{2n}}}{(1+\hbar_1)^{2n}e^{2(a_0+1)\sum_{i=1}^{n}\Delta\tau_i}},$$

we have

$$\|X(t^1_{s,v_1})\| < \frac{e^{-2(1+a_0)\sum_{i=1}^{n}\Delta\tau_i}\delta}{(1+\hbar_1)^{2n}}.$$

It can be shown from Propositions 4.3.3 and 4.3.4 that

$$V(X(t), m(t)) \le (1+\hbar_1)^{2n}e^{2(a_0+1)\sum_{i=1}^{n}\Delta\tau_i}V(X(t^k_{s,v_0}), m(t^k_{s,v_0}));$$
$$t \in [t^k_{s,v_0}, t^{k+1}_{s,v_0}] \cup [0, t_{s,v_0^1}],$$
$$V(X(t^k_{s,v_0}), m(t^{k+1}_{s,v_0})) \le \tilde{q}_0(\xi_0)V(X(t^k_{s,v_0}), m(t^k_{s,v_0})); \ t \in [t^k_{s,v_0}, t^{k+1}_{s,v_0}].$$

From Theorem 2.5.2, the nonlinear switched and impulsive system (4.1) is locally asymptotically stable with respect to $X_e = 0$.   □

Similar to the linear switched and impulsive systems, we have the following result on the stability of nonlinear switched and impulsive systems in the case that the duration time of each CVS is in an interval.

Theorem 4.3.2. *A nonlinear switched and impulsive system (4.1) is locally asymptotically stable with respect to $X_e = 0$ if all $G(j)(j = 1, 2, \cdots, \theta_0)$ defined in equation (3.43) are less than 0*   □

4.4 Stabilization of Nonlinear Switched Systems

We consider the input-to-state stabilization of nonlinear switched systems modelled by

$$\dot{X}(t) = f(X(t), v(t), m(t)), \tag{4.43}$$

where $X(t) \in R^r$ and $v(t) \in R^p$ are, respectively, the continuous state and the control input, the switchings of the system are arbitrary, and $\{f : R^r \times R^p \times \bar{M} \to R^r\}$ is a family of sufficiently regular functions. Each $i \in \bar{M}$ represents a system dynamic that is governed by the corresponding vector field $f(X(t), v(t), i)$ with $f(0, 0, i) = 0$.

A continuous function $\gamma : R_+ \to R_+$ is a $\mathcal{K}$ function if it is strictly increasing and $\gamma(0) = 0$; it is a $\mathcal{K}_\infty$ function if it is a $\mathcal{K}$ function and also $\gamma(r) \to \infty$ as $r \to \infty$. A function $\beta : R_+ \times R_+ \to R_+$ is a $\mathcal{K}\mathcal{L}$ function if for each fixed s the function $\beta(r, s)$ is a $\mathcal{K}$ function with respect to r, and for each fixed r the function $\beta(r, s)$ is decreasing with respect to s and $\beta(r, s) \to 0$ as $s \to \infty$.

Definition 4.4.1. *[177] System (4.43) is said to be input-to-state stable (ISS) if there exist a $\mathcal{K}\mathcal{L}$ function β and a $\mathcal{K}$ function γ such that for any $X(t_0)$ and for any locally essentially bounded input $v(\cdot)$ on $[0, \infty)$ the solution satisfies*

$$\|X(t)\| \leq \beta(\|X(t_0)\|, t - t_0) + \gamma(\|v(t_0, t)\|) \tag{4.44}$$

for all t_0 and t such that $t \geq t_0 \geq 0$.   □

Remark 4.4.1. In inequality (4.44), let $t = t_0$ and $v(t) = 0$, then we have $\|X(t_0)\| \leq \beta(\|X(t_0)\|, 0)$. That is, inequality

$$\beta(s, 0) \geq s \tag{4.45}$$

holds for any $\mathcal{K}\mathcal{L}$ function β satisfying (4.44) and any $s \in R_+$.   □

Definition 4.4.2. *System (4.43) is said to be input-to-state stabilizable if there exists an input $v(t) = K(X(t), u(t), m(t))$ with $u(t)$ being the reference input such that $\dot{X}(t) = \bar{f}(X(t), u(t), m(t)) = f(X(t), K(X(t), u(t), m(t)), m(t))$ is ISS.*   □

Here we also adopt the following assumption, which is standard in the area related to input-to-state stabilization of single nonlinear systems (e.g., see [168, 177] and the references therein).

Assumption 4.3 *For each CVS $i(i \in \bar{M})$ of system (4.43), there exists an input $v(t) = K(X(t), u(t), i)$ such that for any locally essentially bounded input $u()$, we have*

$$\|X(t)\| \le \beta_i(\|X(t_0)\|, t - t_0) + \gamma_i(\|u(t_0, t)\|); \quad t \ge t_0 \ge 0, \tag{4.46}$$

where β_i is a $\mathcal{KL}$ function and γ_i is a $\mathcal{K}$ function.   □

Remark 4.4.2. Condition (4.46) implies that each CVS is input-to-state stabilizable. Note that the input-to-state stabilization of switched systems was also considered in [113]. However, β_i is in a special form of $cX(0)e^{-\lambda t}$ in [113]. When β_i is in a general form as in (4.46), the design and analysis will be much more difficult.   □

The objective here is to derive proper conditions for input-to-state stabilization of system (4.43). To this end, we need the following two supporting results:

Lemma 4.4.1. *Suppose that $\phi_i(i = 1, 2, \cdots, l)$ are $\mathcal{KL}$ functions, $\rho_i(i = 1, 2, \cdots, l)$ are $\mathcal{K}$ functions and $a_i(i = 1, 2, \cdots, k) \in R_+$, then $\sum_{i=1}^{l} \phi_i$ is a $\mathcal{KL}$ function, $\sum_{i=1}^{l} \rho_i$ is a $\mathcal{K}$ function, and*

$$\rho_i(a_1 \bigoplus \cdots \bigoplus a_k) = \rho_i(a_1) \bigoplus \cdots \bigoplus \rho_i(a_k), \tag{4.47}$$

$$\phi_i(a_1 \bigoplus \cdots \bigoplus a_k, s) = \phi_i(a_1, s) \bigoplus \cdots \bigoplus \phi_i(a_k, s). \tag{4.48}$$

Proof. The results can be obtained from the definitions of $\mathcal{K}$ and $\mathcal{KL}$ functions.   □

Lemma 4.4.2. *Assume that $\phi_i(i = 1, 2, \cdots, l)$ are $\mathcal{KL}$ functions. For any positive constants a and $\zeta < 1$, let*

$$\beta(s, t) = \sum_{i,j=1(i \ne j)}^{l} a\phi_i(a\phi_j(a\zeta^{h(t,t_0)}s, 0), 0), \tag{4.49}$$

where $h(t, t_0)$ is an increasing function of t and $h(t, t_0) \to \infty$ as $t \to \infty$. Then $\beta(s, t)$ is also a $\mathcal{KL}$ function.

Proof. For any fixed t, it is clear that $\beta(s,t)$ is a $\mathcal{K}$ function.

For any fixed s, note that $\zeta < 1$ and $h(t,t_0)$ is an increasing function of t. Thus, from (4.49) we know that $\beta(s,t)$ is a decreasing function of t. Note also that $h(t,t_0) \to \infty$ as $t \to \infty$. It follows that $\beta(s,t) \to 0$ as $t \to \infty$. Therefore, $\beta(s,t)$ is a $\mathcal{KL}$ function. □

We will consider both the synchronous case and the asynchronous case in the following sub-sections.

4.4.1 Synchronous Switchings

Let t_k^s denote the kth switching instant of the system (4.43), while t_k^c denote the kth switching instant of the controllers. With synchronous switchings, $t_k^s = t_k^c$.

Since switched systems might become unstable even when all CVSs are stable, a proper switching law of the system is required to guarantee the stability of the switched system. Similarly, we also need such requirements on the switching law of the system to input-to-state stabilize the system (4.43) even when Assumption 4.3 holds.

Let $t_{s,i}^k$ and $t_{f,i}^k$ denote respectively the kth starting time and the kth ending time of CVS i ($i \in \bar{M}$). They are required to satisfy Equation (3.82). Similar to [113], suppose that $\Delta_{1,i}(i = 1, \cdots, n)$ are large enough such that for any $s \in R_+$, we have

$$\beta_i(2\beta_j(2s, \Delta_{1,j}), \Delta_{1,i}) \le \mathrm{L}s < s; \quad \forall i,j \in \bar{M}, \tag{4.50}$$

where $0 < \mathrm{L} < 1$ and β_i ($i \in \bar{M}$) satisfies condition (4.46).

A possible method to verify inequality (4.50) is to calculate the limit

$$\lim_{\Delta_{1,i},\Delta_{1,j} \to \infty} \frac{\beta_i(2\beta_j(2s, \Delta_{1,j}), \Delta_{1,i})}{s} \; ; \; \forall i,j \in \bar{M}.$$

If all the results are less than 1, then inequality (4.50) holds for some large values of $\Delta_{1,i}$ and $\Delta_{1,j}$, based on the definition of limit.

Remark 4.4.3. Under Assumption 4.3 and the above switching law of the system, it can be easily shown that system (4.43) is input-to-state stabilized if the number of switchings is finite. Thus, we only consider the case where the number of switchings is infinite. □

Theorem 4.4.1. *Consider system (4.43) satisfying Assumption 4.3. Suppose that the switchings of the controllers coincide exactly with those of system modes satisfying (4.50). Then, the system is input-to-state stabilized and*

$$\|X(t)\| \le \bar{\beta}(\|X(t_0)\|, t - t_0) + \bar{\gamma}(\|u(t_0, t)\|); \quad t \ge t_0 \ge 0 \tag{4.51}$$

where

$$\bar{\beta}(\|X(t_0)\|, t - t_0) = \sum_{i,j=1(i \ne j)}^{n} \beta_i(2\beta_j(2L^l \|X(t_0)\|, 0), 0),$$

$$\bar{\gamma}(\|u(t_0, t)\|) = \tilde{\gamma}_0 + \gamma_0,$$

$$\tilde{\gamma}_0 = \sum_{i,j=1(i \ne j)}^{n} \beta_i(2\beta_j(2\gamma_0, 0), 0),$$

$$\gamma_0 = \sum_{i=1}^{n} \gamma_i(\|u(t_0, t)\|), \tag{4.52}$$

$$l = r(\frac{k}{2}),$$

and k denotes the total number of switchings of system modes from time t_0 to t.

Proof. For ease of presentation, we let $m_k = m(t_k^c)$, $\gamma_{m_k} = \gamma_{m_k}(\|u(t_k^c, t_{k+1}^c)\|)$. In the following proof we shall use the fact that

$$\Gamma(r_1 + r_2, s) \le \Gamma(2r_1, s) \bigoplus \Gamma(2r_2, s) \tag{4.53}$$

for any $\mathcal{KL}$ function Γ and any nonnegative constants r_1, r_2. From Lemma 4.4.1, Assumption 4.3, condition (4.50), and the property expressed in (4.45), we have

$$\|X(t_1^c)\| \le \beta_{m_0}(\|X(t_0)\|, t_1^c - t_0^c) + \gamma_{m_0}$$
$$\le \beta_{m_0}(\|X(t_0)\|, \Delta T_{m_0}) + \gamma_{m_0};$$
$$\|X(t_2^c)\| \le \beta_{m_1}(\|X(t_1^c)\|, t_2^c - t_1^c) + \gamma_{m_1}$$
$$\le \beta_{m_1}(\beta_{m_0}(\|X(t_0)\|, \Delta T_{m_0}) + \gamma_{m_0}, \Delta T_{m_1}) + \gamma_{m_1}$$
$$\le \beta_{m_1}(2\beta_{m_0}(\|X(t_0)\|, \Delta T_{m_0}), \Delta T_{m_1}) \bigoplus \beta_{m_1}(2\gamma_{m_0}, \Delta T_{m_1}) + \gamma_{m_1}$$
$$\le L\|X(t_0)\| \bigoplus \beta_{m_1}(2\gamma_{m_0}, \Delta T_{m_1}) + \gamma_{m_1};$$
$$\|X(t_3^c)\| \le \beta_{m_2}(\|X(t_2^c)\|, t_3^c - t_2^c) + \gamma_{m_2}$$
$$\le \beta_{m_2}(L\|X(t_0)\| \bigoplus \beta_{m_1}(2\gamma_{m_0}, \Delta T_{m_1}) + \gamma_{m_1}, \Delta T_{m_2}) + \gamma_{m_2}$$
$$\le \beta_{m_2}(2L\|X(t_0)\|, \Delta T_{m_2}) \bigoplus \beta_{m_2}(2\beta_{m_1}(2\gamma_{m_0}, \Delta T_{m_1}), \Delta T_{m_2}) \bigoplus$$
$$\beta_{m_2}(2\gamma_{m_1}, \Delta T_{m_2}) + \gamma_{m_2}$$
$$\le \beta_{m_2}(2L\|X(t_0)\|, \Delta T_{m_2}) \bigoplus L\gamma_{m_0} \bigoplus \beta_{m_2}(2\gamma_{m_1}, \Delta T_{m_2}) + \gamma_{m_2}.$$

Using the inductive method, we can get the following inequalities when the number of switchings of system modes k is even and $k > 3$,

$$\|X(t_k^c)\| \le \beta_{m_{k-1}}(\|X(t_{k-1}^c)\|, t_k^c - t_{k-1}^c) + \gamma_{m_{k-1}}$$
$$\le L^l\|X(t_0)\| \bigoplus \beta_{m_{k-1}}(2L^{l-1}\gamma_{m_0}, \Delta T_{m_{k-1}}) \bigoplus L^{l-1}\gamma_{m_1}$$
$$\bigoplus \beta_{m_{k-1}}(2L^{l-2}\gamma_{m_2}, \Delta T_{m_{k-1}}) \bigoplus L^{l-2}\gamma_{m_3} \bigoplus \cdots$$
$$\bigoplus \beta_{m_{k-1}}(2L\gamma_{m_{k-4}}, \Delta T_{m_{k-1}}) \bigoplus L\gamma_{m_{k-3}}$$
$$\bigoplus \beta_{m_{k-1}}(2\gamma_{m_{k-2}}, \Delta T_{m_{k-1}}) + \gamma_{m_{k-1}},$$

where $l = r(\frac{k}{2})$.

Thus, for any $t \in [t_k^c, t_{k+1}^c]$,

$$\|X(t)\| \le \beta_{m_k}(\|X(t_k^c)\|, t - t_k^c) + \gamma_{m_k}(\|u(t_k^c, t)\|)$$
$$\le \beta_{m_k}(2L^l\|X(t_0)\|, t - t_k^c) \bigoplus \beta_{m_k}(2\beta_{m_{k-1}}(2L^{l-1}\gamma_{m_0}, \Delta T_{m_{k-1}}), t - t_k^c)$$
$$\bigoplus \beta_{m_k}(2L^{l-1}\gamma_{m_1}, t - t_k^c) \bigoplus \beta_{m_k}(2\beta_{m_{k-1}}(2L^{l-2}\gamma_{m_2}, \Delta T_{m_{k-1}}), t - t_k^c)$$
$$\bigoplus \beta_{m_k}(2L^{l-2}\gamma_{m_3}, t - t_k^c) \bigoplus \cdots$$
$$\bigoplus \beta_{m_k}(2\beta_{m_{k-1}}(2L\gamma_{m_{k-4}}, \Delta T_{m_{k-1}}), t - t_k^c)$$
$$\bigoplus \beta_{m_k}(2L\gamma_{m_{k-3}}, t - t_k^c) \bigoplus \beta_{m_k}(2\beta_{k-1}(2\gamma_{m_{k-2}}, \Delta T_{m_{k-1}}), t - t_k^c)$$
$$\bigoplus \beta_{m_k}(2\gamma_{m_{k-1}}, t - t_k^c) + \gamma_{m_k}(\|u(t_k^c, t)\|).$$

From (4.52), we replace $\gamma_{m_i}, i = 0, \cdots, k$, with γ_0, and notice that for any $a, b \in R_+$, $a \bigoplus b \le a + b$ and $a \bigoplus b = a$ if $a \ge b$. Thus, we can further obtain

$$\|X(t)\| \le \beta_{m_k}(2L^l\|X(t_0)\|, 0) \bigoplus \beta_{m_k}(2\beta_{m_{k-1}}(2\gamma_0, 0), 0) + \gamma_0$$
$$\le \beta_{m_k}(2L^l\|X(t_0)\|, 0) + \beta_{m_k}(2\beta_{m_{k-1}}(2\gamma_0, 0), 0) + \gamma_0$$
$$\le \bar{\beta}(\|X(t_0)\|, t - t_0) + \tilde{\gamma} + \gamma_0 \qquad (4.54)$$
$$= \bar{\beta}(\|X(t_0)\|, t - t_0) + \bar{\gamma}(\|u(t_0, t)\|).$$

In a similar way, it can be shown that (4.54) holds in the case where k is odd and $k > 3$.

Note that $l = r(k/2)$ is an increasing function of t and $l \to \infty$ as $t \to \infty$. From Lemma 4.4.2, $\bar{\beta}(\|X(t_0)\|, t - t_0)$ is a $\mathcal{KL}$ function. Therefore, system (4.43) is input-to-state stabilized in this case. $\square$

Remark 4.4.4. It should be emphasized that we have used inequality (4.53) in the above proof. Note that the whole derivation cannot proceed if we employ the fact

$$\Gamma(r_1 + r_2, s) \le \Gamma(2r_1, s) + \Gamma(2r_2, s),$$

which is usually used when the input-to-state stabilization of single nonlinear system is studied. $\square$

Let us use a numerical example to illustrate the application of the proposed stabilization result.

Example 4.4.1. Consider a nonlinear switched system consisting of the following two first-order CVSs:

$$\text{CVS 1:} \quad \dot{X}(t) = X^3(t) - \frac{X^3(t)v^2(t)}{2},$$

$$\text{CVS 2:} \quad \dot{X}(t) = X(t) + v(t),$$

where $X(t) \in R$ and $v(t) \in R$.

It can be shown that Assumption 4.3 holds with $v(t) = \sqrt{2 + u^2(t)}$, $\beta_1(r, s) = \frac{r}{\sqrt{2r^2 s + 1}}$ and $\gamma_1(s) = s$ for CVS 1, and $v(t) = -2X(t) + u(t)$, $\beta_2(r, s) = re^{-s}$ and $\gamma_2(s) = s$ for CVS 2. Moreover, it can be checked that

$$\lim_{\Delta_{1,1}, \Delta_{1,2} \to \infty} \frac{\beta_1(2\beta_2(2s, \Delta_{1,2}), \Delta_{1,1})}{s} = \lim_{\Delta_{1,1}, \Delta_{1,2} \to \infty} \frac{\beta_2(2\beta_1(2s, \Delta_{1,1}), \Delta_{1,2})}{s} = 0.$$

Thus, (4.50) holds for some large $\Delta_{1,1}$ and $\Delta_{1,2}$. For example, if $\Delta_{1,1} = \Delta_{1,2} = 2s$, then $L = 4e^{-2}$ and (4.50) holds. Also note that the results in [113] cannot be used to study this example.

Using Theorem 4.4.1, we know that

$$\|X(t)\| \le 8L^l \|X(t_0)\| + 16\|u(t_0, t)\|; \quad t \ge t_0 \ge 0,$$

where l is defined in Theorem 4.4.1. For our simulation studies, we take the switching instant of the system as the values shown in Fig. 4.1, CVS 1 as the initial mode, and we let $u(t) = 3sin(t)$ and $X(0) = 3$. The simulation result, illustrated in Figs. 4.2 and 4.3 with $\beta + \gamma$ standing for $8L^l \|X(t_0)\| + 16\|u(t_0, t)\|$, indicates that the considered system is input-to-state stabilized. $\square$

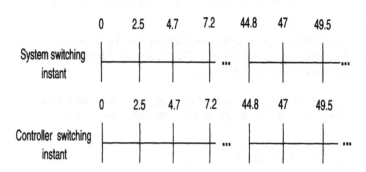

Fig. 4.1. The switchings of system and controller

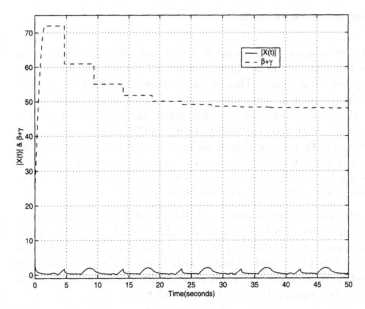

Fig. 4.2. Simulation results in the synchronous case

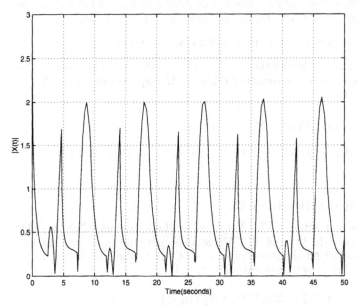

Fig. 4.3. Simulation results in the synchronous case

4.4.2 Asynchronous Switchings

In practice, the switchings of the controller may not coincide exactly with those of the system since we do not know the initial CVS and the subsequent CVSs of the system in advance. Thus, it is necessary to identify them and then switch from the present controller to the corresponding controllers. As expected, the design and analysis are now much more involved than the synchronous case, since we need to identify the initial CVS and the subsequent CVSs of the system. To achieve this, a delay is imposed on the switchings of the subcontrollers, that is, $t_k^c > t_k^s$ $(k = 0, 1, 2, \cdots)$. Intervals $[t_k, t_k^c]$ $(k = 0, 1, 2, \cdots)$ are used to do the identification. Once the active CVS is known, the corresponding sub-controller is switched into action.

Similar to [60], a model-based scheme, as illustrated in Fig. 3.9, is adopted to identify the subsequent CVSs. Assume that there is only one CVS model whose state is equal to the state of system (4.43) for any control input and any interval if system (4.43) and all the models of the CVSs have the same initial state and there is no measurement noise or disturbance. Without loss of generality, we also suppose that $X(t) \neq 0$ for all $t \geq t_0$.

Since t_k^s is unknown, we also need to estimate it. Thus, the whole task is composed of two steps: estimate the kth switching instant of the system and identify the kth active CVS. These are given in details as follows:

Step 1: Estimate the kth switching instant of the system.

In Fig. 3.9, t_k^{se} and $\tilde{X}_i$ denote the estimate of the kth switching instant of the system and the state of the model of CVS $i \in \bar{M}$ respectively. Then, $t_0^{se} = t_0$ and $t_k^{se}(k \geq 1)$ are

$$t_k^{se} = \sup_t \{t > t_{k-1}^c | X(t) = \tilde{X}_{m(t_{k-1}^c)}(t) \text{ and}$$

$$\|X(t)\| \leq \beta_{m(t_{k-1}^c)}(\|X(t_{k-1}^c)\|, t - t_{k-1}^c) + \gamma_{m(t_{k-1}^c)}(\|u(t_{k-1}^c, t)\|)\}. \qquad (4.55)$$

Step 2: Identify the kth active CVS.

To identify the kth active CVS, $X(t_k^{se})$ is feedback to each CVS model to ensure that system (4.43) and all CVS models have the same state at time point t_k^{se}. To avoid that the states of system (4.43) escape into infinity before a proper controller is switched into action, t_k^c is

$$t_k^c = \sup_t \{t_k^{se} \leq t \leq t_k^e + \Delta t \,|\, \|X(t)\| \leq a\|X(t_k^e)\|\}, \qquad (4.56)$$

where $t_k^{se} = t_0$, $a > 1$ and $\Delta t = \min_{1 \leq i \leq n} \frac{\Delta_{1,i} - \Delta T_i'}{2}$. Here, a and $\Delta T_i'$ $(i = 1, \cdots, n)$ are determined by letting

$$H(p, t_i, t_j) = p\beta_i(2p\beta_j(2s, t_j), t_i); \quad p, t_i, t_j \in R_+, i, j \in \bar{M}.$$

Obviously, $H(p, t_i, t_j)$ is a continuous function of p, t_i and t_j. From condition (4.50), we have $H(1, \Delta_{1,i}, \Delta_{1,j}) \leq \bar{L}s < s$. It follows that there exist $a > 1$ and $\bar{L}(0 < \bar{L} < 1)$, $\Delta T_i' < \Delta_{1,i}$ and $\Delta T_j' < \Delta_{1,j}$ such that

$$H(a, \Delta T_i', \Delta T_j') \leq \bar{L}s < s. \tag{4.57}$$

Thus, the present active CVS can be obtained from the state of system (4.43) and all the models of the CVSs within $[t_k^e, t_k^c]$. Based on the above discussion, we can also show that system (4.43) under $v(t) = K(X(t), u(t), m(t))$ ($m(t) = 1, \cdots, n$) is ISS in the asynchronous case, and

$$\|X(t)\| \leq \hat{\beta}(\|X(t_0)\|, t - t_0) + \hat{\gamma}(\|u(t_0, t)\|), \tag{4.58}$$

where

$$\hat{\beta}(\|X(t_0)\|, t - t_0) = \sum_{i,j=1(i\neq j)}^{n} a\beta_i(2a\beta_j(2\bar{L}^{\lfloor k/2 \rfloor}a\|X(t_0)\|, 0), 0),$$

$$\hat{\gamma}(\|u(t_0, t)\|) = \sum_{i,j=1(i\neq j)}^{n} a\beta_i(2a\beta_j(2\gamma_0), 0), 0) + \sum_{i=1}^{n} a\gamma_i(\|u(t_0, t)\|),$$

and k denotes the total number of switchings of system modes from t_0 to t.

As an illustration of input-to-state stabilization with asynchronous switchings, we present the following example.

Example 4.4.2. Consider the nonlinear switched system given in Example 4.4.1. From (4.58), we can determine the $\mathcal{KL}$ function $\hat{\beta}(\|X(t_0)\|, t - t_0) = 8\bar{L}^l a^3 \|X(t_0)\|$ and the $\mathcal{K}$ function $\hat{\gamma}(\|u(t_0, t)\|) = (16a^3 + 2a)\|u(t_0, t)\|$. That is, $\|X(t)\| \leq 8\bar{L}^l a^3 \|X(t_0)\| + (16a^3 + 2a)\|u(t_0, t)\|$; $t \geq t_0 \geq 0$.

From (4.57), it can be computed that $a = 1.2$, $\bar{L} = 5.76e^{-1.8}$ and $\Delta T_1' = \Delta T_2' = 1.8s$. As a simulation example, the estimates of the switching instant

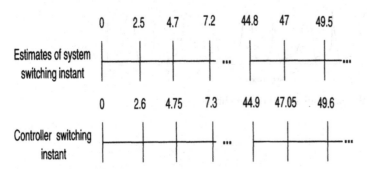

Fig. 4.4. The switchings of system and controller

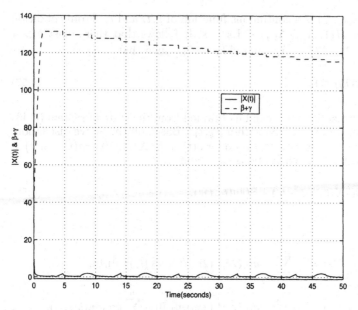

Fig. 4.5. Simulation results in the asynchronous case

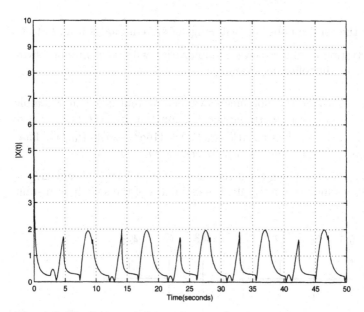

Fig. 4.6. Simulation results in the asynchronous case

of the system and the switching instant of controllers are given in Fig. 4.4. The results shown in Figs. 4.5 and 4.6 also demonstrate that the system with the same initial conditions as above is input-to-state stabilized in the asynchronous case. In the figure, $\beta + \gamma$ stands for $8\bar{L}^l a^3 \|X(t_0)\| + (16a^3 + 2a)\|u(t_0, t)\|$.   □

5. Impulsive Synchronization of Chaotic Systems

In this chapter, less conservative conditions on the stability of impulsive systems are first derived. These results are then used to study the stabilization and synchronization of a class of chaotic systems via impulsive control.

5.1 Chaotic Systems

A chaotic system is a deterministic system that exhibits random behavior through its sensitive dependence on initial conditions. The behavior of a chaotic system is unpredictable and, therefore, random-like. There is no universally agreed definition of chaos. For engineering applications, there are at least three basic dynamical properties that characterize a chaotic behavior [120]:

1. An essentially continuous, and banded frequency spectrum that resembles random noise.

2. Extremely sensitive to initial conditions, that is, nearby orbits diverge very rapidly;

3. An ergodicity and mixing of the dynamical orbits, which implies the visit of the entire phase space by the chaotic behavior.

Example 5.1.1. The Chua's circuit is described by equation (1.16). Let the parameters of Chua's circuit be $k = 1$, $\vartheta_1 = -0.722451121$, $\vartheta_2 = -1.138411196$, $\alpha = -9.35159085$, $\beta = -14.790313805$ and $\gamma = -0.016073965$. The state trajectories are illustrated in Figs. 5.1 and 5.2. Clearly, the Chua's circuits have the above three properties. For example, Chua's circuit is extremely sensitive to small differences in initial conditions, as shown by the state trajectories from two different initial values $[-0.2, -0.02, 0.1]$ and $[-0.8, -0.06, 0.1]$ in Figs. 5.3-5.5. $\qquad\square$

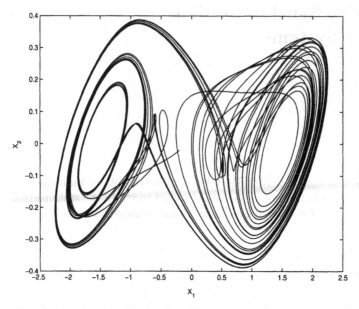

Fig. 5.1. The state trajectories of $[x_1, x_2]$

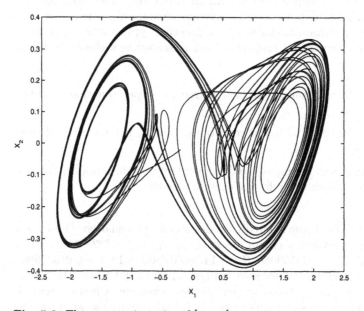

Fig. 5.2. The state trajectories of $[x_1, x_3]$

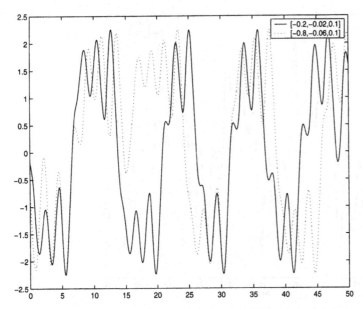

Fig. 5.3. The state trajectories of x_1 with different initial values

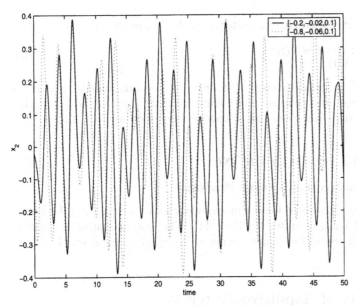

Fig. 5.4. The state trajectories of x_2 with different values

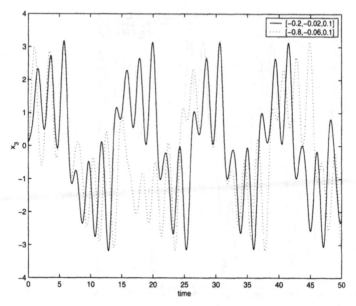

Fig. 5.5. The state trajectories of x_3 with different values

The commonly used chaotic systems can be completely described by

$$\begin{cases} \dot{X}(t) = AX(t) + \phi(X(t)) \\ Y(t) = CX(t) \end{cases}, \tag{5.1}$$

where $X \in R^r$ is a state vector, A is an $r \times r$ constant matrix, $y \in R^m$ is an output vector, and C is an $m \times r$ constant matrix, and $\phi : R^r \to R^r$ is a nonlinear function that gives rise to chaotic dynamics, and it satisfies

$$2X^T(t)\phi(X(t)) \leq X^T(t)(H + H^T)X(t). \tag{5.2}$$

where H is a matrix that is known in advance.

Two frequently studied lower-dimensional chaotic systems are the Lorenz systems and Chua's circuits [48, 120, 196]. The Lorenz system is given by equation (1.20) while the Chua's circuit is described by equation (1.16).

5.2 Stability of Impulsive Systems

The chaotic system of (5.1) when subject to impulsive jumps in states will become an impulsive system described by [85]

$$\begin{cases} \dot{X}(t) = AX(t) + \phi(X(t)) \; ; \; t \neq t_k \\ \Delta X(t) \triangleq X(t^+) - X(t_k^-) = I_k(X) \; ; \; t = t_k, k = 1, 2 \cdots \end{cases}, \tag{5.3}$$

where $f : R_+ \times R^r \to R^r (R_+ = [0, \infty))$ and $I_k : R^r \to R^r$ are continuous, $X \in R^r$ is the state variable, $\{t_k\}(k = 1, 2, \cdots)$ satisfy

$$0 < t_1 < t_2 < \cdots < t_k < t_{k+1} < \cdots ; \ t_k \to \infty \ \text{as} \ k \to \infty.$$

Instead of using individual CVS as a unit, we have chosen a cycle as the unit for stability analysis of switched and impulsive systems in chapters 2-4. The Lyapunov function is only required to be non-increasing along the cycle unit, and the condition is thus relaxed. Similar idea is used in this section, where a bigger unit is adopted to derive less conservative condition on the stability of impulsive systems. Specifically, the unit is selected as the chaotic system and the impulsive control within two successive impulsive intervals instead of those within one interval. The Lyapunov function is also only required to be non-increasing along the bigger unit.

To establish sufficient conditions for the stability of impulsive systems, we need the definition of comparison system, which plays an important role in the stability analysis of impulsive systems.

Definition 5.2.1. *[85] Let V be a Lyapunov function and assume that*

$$D^+V(t, X) \leq g(t, V(t, X)) \ ; \ t \neq t_k, \tag{5.4}$$
$$V(t, X + \Delta X) \leq \psi_k(V(t, X)) \ ; \ t = t_k, \tag{5.5}$$

where $g : R_+ \times R_+ \to R$ *is continuous,* v_0 *is a set defined in [85] and* $\psi_k : R_+ \to R_+$ *is nondecreasing. Then the system*

$$\begin{cases} \dot{w} = g(t, w) \ ; \ t \neq t_k \\ w(t_k^+) = \psi_k(w(t_k)) \\ w(t_0^+) = w_0 \geq 0 \end{cases} \tag{5.6}$$

is called the comparison system of (5.3). □

Let $g(t, w) = \dot{\lambda}(t)w$, $\lambda \in C^1[R_+, R_+]$, $\psi_k(w) = d_k w$, $d_k > 0$ for all $k \in \{1, 2, \cdots\}$. Then, we have the following stability result:

Theorem 5.2.1. *The origin of system (5.3) is asymptotically stable if the following conditions hold:*

1. $d_0 = \inf_k\{d_k\}$ *exists and* $d_0 > 0$.

2. There exists an $L(0 < L < 1)$ *such that*

$$\lambda(t_{2k+3}) + \ln(\frac{d_{2k+2}d_{2k+1}}{L^2}) \leq \lambda(t_{2k+1}) \ \text{for} \ k = 0, 1, \cdots \tag{5.7}$$

3. $\lambda(t)$ *satisfies that*

$$\dot{\lambda}(t) \geq 0. \tag{5.8}$$

4. *There exist $\alpha(\cdot)$ and $\beta(\cdot)$ in Class K [85] such that*

$$\beta(\|X\|) \leq V(t, X) \leq \alpha(\|X\|). \tag{5.9}$$

Proof. It can be seen that the solution $w(t, t_0, w_0)$ of the comparison system

$$\begin{cases} \dot{w}(t) = \dot{\lambda}(t)w(t) \\ w(t_k^+) = d_k w(t_k) \\ w(t_0^+) = w_0 \geq 0 \end{cases} \tag{5.10}$$

is

$$w(t, t_0, w_0) = w_0 \prod_{t_0 < t_k < t} d_k \exp(\lambda(t) - \lambda(t_0)). \tag{5.11}$$

We shall show that

$$w(t, t_0, w_0) \leq \frac{w_0}{\min\{1, d_0\}} \exp(\lambda(t_1) - \lambda(t_0)) \; ; \; t \geq t_0 \; ; \; 0 \leq t_0 < t_1. \tag{5.12}$$

To do this, the following three cases are considered.

Case 1. If $t_0 < t < \tau_1$, then

$$\begin{aligned} w(t, t_0, w_0) &= w_0 \exp(\lambda(t) - \lambda(t_0)) \\ &\leq w_0 \exp(\lambda(t_1) - \lambda(t_0)) \\ &\leq \frac{w_0}{\min\{1, d_0\}} \exp(\lambda(t_1) - \lambda(t_0)). \end{aligned}$$

Case 2. If $t_{2k-1} < t < t_{2k}$, then for all k greater than 0,

$$\begin{aligned} w(t, t_0, w_0) &= w_0 \prod_{i=1}^{2k-1} d_i \exp(\lambda(t_1) - \lambda(t_0)) \exp(\lambda(t) - \lambda(t_1)) \\ &\leq w_0 \frac{1}{d_0} \prod_{i=1}^{2k} d_i \exp(\lambda(t_1) - \lambda(t_0)) \exp(\lambda(t_{2k+1}) - \lambda(t_1)) \\ &\leq \frac{w_0}{d_0 r^k} \exp(\lambda(t_1) - \lambda(t_0)) \\ &\leq \frac{w_0}{\min\{1, d_0\}} \exp(\lambda(t_1) - \lambda(t_0)). \end{aligned}$$

Case 3. If $t_{2k} < t < t_{2k+1}$, then for all k greater than 0,

$$\begin{aligned} w(t, t_0, w_0) &= w_0 \prod_{i=1}^{2k} d_i \exp(\lambda(t_1) - \lambda(t_0)) \exp(\lambda(t) - \lambda(t_1)) \\ &\leq w_0 \prod_{i=1}^{2k} d_i \exp(\lambda(t_1) - \lambda(t_0)) \exp(\lambda(t_{2k+1}) - \lambda(t_1)) \\ &\leq \frac{w_0}{r^k} \exp(\lambda(t_1) - \lambda(t_0)) \\ &\leq \frac{w_0}{\min\{1, d_0\}} \exp(\lambda(t_1) - \lambda(t_0)). \end{aligned}$$

Therefore, (5.12) holds.

Note that $k \to \infty$ as $t \to \infty$. From Case 2 and Case 3, we know that

$$\lim_{t \to \infty} w(t, t_0, w_0) = 0.$$

Similar to the proof of Theorem 3.2.1 in [85], it can be shown that the origin of (5.3) is asymptotically stable. □

Remark 5.2.1. It can be derived from inequality (5.7) that

$$V(t_{2k+3}, X(t_{2k+3})) \le L^2 V(t_{2k+1}, X(t_{2k+1})) \; ; \; \forall k \tag{5.13}$$

Thus, the Lyapunov function is non-increasing along the bigger unit. □

Remark 5.2.2. (5.7) can be generalized to the following condition:

There exist a finite integer $m_0 > 0$ and an $0 < L < 1$ such that

$$\lambda(t_{m_0(k+1)+1}) + \ln\left(\frac{d_{m_0(k+1)} \cdots d_{m_0 k+1}}{L^{m_0}}\right) \le \lambda(t_{m_0 k+1}) \; ; \; k = 0, 1, \cdots \tag{5.14}$$

Similar to the choice of a Lyapunov function, the choice of m_0 in (5.14) depends on the actual system considered. □

5.3 Impulsive Control of Chaotic Systems

In this section, the stability results obtained in the previous section is used to design impulsive control for the chaotic systems (5.1). The control time instant is given by

$$0 < t_1 < t_2 < \cdots < t_k < t_{k+1} < \cdots \; ; \; t_k \to \infty \text{ as } k \to \infty,$$

and the control $U(t_k, X)$ is

$$U(t_k, X(t_k)) = BY(t_k). \tag{5.15}$$

We can obtain a nonlinear impulsive system as

$$\begin{cases} \dot{X}(t) = AX(t) + \phi(X(t)) \\ Y(t) = CX(t) \; ; \; t \ne t_k \\ X(t_k^+) = X(t_k^-) + U(t_k^-, X(t_k^-)) \\ U(t_k^-, X(t_k^-)) = BCX(t_k^-) \; ; \; k = 1, 2, \cdots \end{cases} \tag{5.16}$$

To use the results obtained in the previous section, the above system is rewritten as

$$\begin{cases} \dot{X}(t) = AX(t) + \phi(X(t)) \; ; \; t \ne t_k \\ \Delta X(t_k) = U(t_k^-, X(t_k^-)) = BCX(t_k^-) \; ; \; k = 1, 2, \cdots . \\ X(t_0^+) \hat{=} X_0 \end{cases} \tag{5.17}$$

Then, we can obtain the following result on the design of impulsive control.

Theorem 5.3.1. *Assume that the $n \times n$ matrix Γ is symmetric and positive definite, and λ_{min} and λ_{max} are respectively the smallest and the largest eigenvalues of Γ. Let*

$$Q = \Gamma A + A^T \Gamma, \tag{5.18}$$

and $Q \leq \gamma_1 \Gamma$ with γ_1 being a constant. Then the origin of impulsive system (5.17) is asymptotically stable if the following conditions hold:

1. *Three inequalities*

    $$\|I + BC\|_2 \leq 1, \tag{5.19}$$
    $$(I + (BC)^T)\Gamma(I + BC) \leq \gamma_2 \Gamma, \tag{5.20}$$
    $$\gamma_1 + \frac{2L\lambda_{max}}{\lambda_{min}} \geq 0, \tag{5.21}$$

    hold with I being the identity matrix and γ_2 a positive constant.

2. *There exists an $0 < L < 1$ such that*

    $$(\gamma_1 + \frac{2L\lambda_{max}}{\lambda_{min}})(t_{2k+3} - t_{2k+1}) \leq -2\ln(\frac{\gamma_2}{L}) \ ; \ k = 0, 1, \cdots . \tag{5.22}$$

Proof. Let

$$V(X) = X^T \Gamma X.$$

It can be easily shown that

$$D^+ V(x) \leq (\gamma_1 + \frac{2L\lambda_{max}}{\lambda_{min}})V(x) \ ; \ t \neq t_k,$$

and

$$V(X + U(t_k, X)) \leq \gamma_2 V(X) \ ; \ t = t_k.$$

Then the comparison system of (5.17)

$$\begin{cases} \dot{w}(t) = (\gamma_1 + \frac{2L\lambda_{max}}{\lambda_{min}})w(t) \ ; \ t \neq t_k \\ w(t_k^+) = \gamma_2 w(t_k) \\ w(t_0^+) = w_0 \geq 0 \end{cases} \tag{5.23}$$

is obtained.

Note that $d_k = \gamma_2$, $\inf_k\{d_k\} = \gamma_2 > 0$, and

$$\lambda(t_{2k+3}) - \lambda(t_{2k+1}) = (\gamma_1 + \frac{2L\lambda_{max}}{\lambda_{min}})(t_{2k+3} - t_{2k+1}).$$

From Theorem 5.2.1, we know that the result holds. $\qquad \square$

Remark 5.3.1. We do not require that BC is symmetric. Thus, our result can be used for a wider class of nonlinear systems when compared with [206]. □

Remark 5.3.2. Matrix B can be obtained by solving the following convex optimization problem:

$$\min\{a\},$$

subject to

$$\begin{bmatrix} a\Gamma & \Upsilon(I+BC) \\ (I+BC)^T\Upsilon^T & I \end{bmatrix} \geq 0, \tag{5.24}$$

$$\begin{bmatrix} I & I+BC \\ (I+BC)^T & I \end{bmatrix} \geq 0, \tag{5.25}$$

where $\Upsilon^T\Upsilon = \Gamma$.

The above problem can be computationally tractable with existing tools, such as the linear matrix inequalities (LMI) [16]. □

Remark 5.3.3. Note that when

$$(\gamma_1 + \frac{2L\lambda_{max}}{\lambda_{min}})(t_{k+1} - t_k) \leq -\ln(\frac{\gamma_2}{L}), 0 < L < 1 \tag{5.26}$$

holds for all k and an $0 < L < 1$ [206], we can obtain (5.22) as follows:

$$(\gamma_1 + \frac{2L\lambda_{max}}{\lambda_{min}})(t_{2k+3} - t_{2k+1})$$
$$= (\gamma_1 + \frac{2L\lambda_{max}}{\lambda_{min}})(t_{2k+3} - t_{2k+2}) + (\gamma_1 + \frac{2L\lambda_{max}}{\lambda_{min}})(t_{2k+2} - t_{2k+1})$$
$$\leq -2\ln(\frac{\gamma_2}{L}).$$

However, (5.26) cannot be derived from (5.22) as (5.22) is only required along a subsequence of τ_k. Thus, Theorem 5.3.1 is less conservative than Theorem 3 in [206]. □

Remark 5.3.4. Condition (5.22) can be generalized to (5.27) given below.

There exist a finite integer $m_0 > 0$ and an $0 < L < 1$ such that

$$(\lambda_3 + \frac{2L\lambda_{max}}{\lambda_{min}})(t_{m_0(k+1)+1} - t_{m_0k+1}) \leq -m_0\ln(\frac{\gamma_2}{L}) ; k = 0,1,\cdots. \tag{5.27}$$

Similar to the choice of a Lyapunov function for a particular system, the choice of m_0 is related to the actual system considered. □

5.4 Synchronization of Chua's Circuits via Impulsive Control

In this section, we shall first use the results of the previous section to consider the stabilization of Chua's circuit (1.16). The matrices A, $\phi(X)$, and H are

$$A = \begin{bmatrix} -\alpha & \alpha & 0 \\ 1 & -1 & 1 \\ 0 & -\beta & -\gamma \end{bmatrix}, \tag{5.28}$$

$$\phi(X) = \begin{bmatrix} -\alpha f(X) \\ 0 \\ 0 \end{bmatrix}, \tag{5.29}$$

$$H = \begin{bmatrix} |\vartheta_2 \alpha| & 0 & 0 \\ 0 & 0 & 0 \\ 0 & 0 & 0 \end{bmatrix}. \tag{5.30}$$

Define a function $\chi(A, H)$ as

$$\chi(A, H) = \lambda_{max}(A + H + (A + H)^T). \tag{5.31}$$

Introduce the following impulsive control:

$$U(t_k, X(t_k)) = BX(t_k) \; ; \; k = 1, 2, \cdots \tag{5.32}$$

where B is a symmetric matrix satisfying $\rho(I + B) < 1$ and $\rho(I + B)$ denotes the spectral radius of matrix $(I + B)$, $t_k(k = 1, 2, \cdots)$ are time varying and satisfy

$$t_{2k+1} - t_{2k} = \epsilon_1(t_{2k} - t_{2k-1}). \tag{5.33}$$

In equation (5.33), ϵ_1 is a given positive constant. That is, in our impulsive control strategy, we consider a pair of impulses as a unit of control signal. In this way, we only need to consider a subsequence of the impulses in our stability analysis. Denote

$$\Delta_1 = \sup\{t_{2k} - t_{2k-1}\} < \infty, \tag{5.34}$$
$$\Delta_2 = \sup\{t_{2k+1} - t_{2k}\} < \infty. \tag{5.35}$$

Then, we have

Corollary 5.4.1. *The origin of Chua's oscillator (1.16) under impulsive control (5.32) and (5.33) is asymptotically stable if*

$$0 \leq \chi(A, H) \leq -\frac{2}{(1 + \epsilon_1)\Delta_1} \ln(\frac{d_1}{L}), \tag{5.36}$$

where $0 < L < 1$ and $d_1 = \rho^2(I + B) < 1$, and $d_1 < L$. □

We shall now study the impulsive synchronization of two Chua's circuits, which are called the driven system and the driving system, respectively [204]. In an impulsive synchronization configuration, the driving system is given by (1.16), whereas the driven system is

$$\dot{\tilde{X}} = A\tilde{X} + \psi(\tilde{X}), \tag{5.37}$$

where $\tilde{X} = (\tilde{x}, \tilde{y}, \tilde{z})^T$ is its state variables, A and ψ are defined in (5.28) and (5.29).

At discrete instant $t_k(k = 1, 2, \cdots)$ defined in (5.33), the state variables of the driving system are transmitted to the driven system and then the state variables of the driven system are subject to jumps at t_k. In this sense, the driven system is

$$\begin{cases} \dot{\tilde{X}} = A\tilde{X} + \psi(\tilde{X}) \; ; \; t \neq t_k \\ \Delta\tilde{X}|_{t=t_k} = -Be \; ; \; k = 1, 2, \cdots \end{cases}, \tag{5.38}$$

where $e^T = (e_x, e_y, e_z) = (x - \tilde{x}, y - \tilde{y}, z - \tilde{z})$ is the synchronization error. Let

$$\Psi(X, \tilde{X}) = \psi(X) - \psi(\tilde{X}) = \begin{bmatrix} -\alpha f(x) + \alpha f(\tilde{x}) \\ 0 \\ 0 \end{bmatrix}. \tag{5.39}$$

The error system of the impulsive synchronization is then given by

$$\begin{cases} \dot{e} = Ae + \Psi(X, \tilde{X}) \; ; \; t \neq t_k \\ \Delta e|_{t=t_k} = Be \; ; \; k = 1, 2, \cdots \end{cases}. \tag{5.40}$$

Similar to the stabilization of Chua's circuits, we can obtain the following result:

Corollary 5.4.2. *The impulsive synchronization of two Chua's circuits, with the error system given by (5.40), is asymptotically stable if (5.36) holds.* □

Remark 5.4.1. Consider the case that $\epsilon_1 < 1$. For any $0 < L < 1$ satisfying $d_1 \leq L$, which is required by [204], we choose that

$$\Delta_1 = -\frac{2ln(\frac{d_1}{L})}{(1+\epsilon_1)\chi(A, H)}. \tag{5.41}$$

It can be shown that (5.36) holds. Thus, the origin of Chua's circuit under impulsive control (5.32) and (5.33) is asymptotically stable. Note that Δ_1 is greater than both the upper bound Δ_{max} in [204] and the upper bound defined in [106]. Thus, a larger bound can be obtained with the proposed approach here. □

In the following, an estimate of the synchronization time is derived for the class of chaotic systems with the given parameters and the impulsive control law. Let $e(t_{2k}, t_0, e_0)$ denote the synchronization error at time t_{2k} for the given initial time t_0 and the initial error e_0. We then have the following result on the synchronization time.

Lemma 5.4.1. *For any $\eta > 0$, let n_0 be*

$$n_0(\eta) = \log_{L^2} \frac{\eta^2}{\|e_0\|^2}. \tag{5.42}$$

Then with the impulsive control satisfying (5.32) and (5.33), when $n \geq n_0(\eta)$,

$$\|e(t_{2n}, t_0, e_0)\| < \eta.$$

Proof. Define the Lyapunov function as

$$V(e) = \|e\|^2.$$

It follows that

$$\dot{V}(e) \leq \lambda_{max}(A + H + (A + H)^T)V(e) \; ; \; t \neq t_k,$$
$$V(e(t_k^+)) \leq d_1 V(e(t_k^-)).$$

For simplicity, we denote

$$\varsigma(t) = \lambda_{max}(A + H + (A + H)^T)t.$$

The bound for the synchronization time can then be computed as follows:

$$
\begin{aligned}
V(e(t_{2n}, t_0, e_0)) &\leq e^{\varsigma(t_{2n}) - \varsigma(t_{2n-1})} V(e(t_{2n-1}^+, t_0, e_0)) \\
&\leq d_1 e^{\varsigma(t_{2n}) - \varsigma(t_{2n-1})} V(e(t_{2n-1}^-, t_0, e_0)) \\
&\leq d_1^2 e^{\varsigma(t_{2n}) - \varsigma(t_{2n-2})} V(e(t_{2n-2}^+, t_0, e_0)) \\
&\leq L^2 V(e(t_{2n-2}^+, t_0, e_0)) \\
&\leq L^{2n} V(e_0).
\end{aligned}
$$

Clearly, when $n \geq n_0(\eta)$, we have $\|e(t_{2n}, t_0, e_0)\| < \eta$. □

Let T_1 and T_2 be the actual time intervals set in the system. Note that T_1 and T_2 cannot be beyond the upper bounds of the impulsive intervals Δ_1 and Δ_2. An estimate of synchronization time required is then given by

$$T_{syn} = n(T_1 + T_2). \tag{5.43}$$

In other words, the chaotic system in the receiver will be synchronized with that in the transmitter after $t \geq T_{syn}$. Two numerical examples are used to illustrate the effectiveness of the obtained results.

Example 5.4.1. Consider the Chua's circuit given in Example 5.1.1. It can be easily computed that

$$\lambda_{max}(A + A^T) + 2|\vartheta_2 \alpha| = 34.6988,$$
$$\chi(A, H) = 16.8385.$$

Choose the control matrix B as

$$B = \begin{bmatrix} l & 0 & 0 \\ 0 & -1 & 0 \\ 0 & 0 & -1 \end{bmatrix},$$

where l is a constant satisfying $-2 < l < 0$. It follows that $d_1 = (l+1)^2$. Consider a special case with $l = -1.05$, then $d_1 = 0.0025$. We set $L = 1/300$ and $\epsilon_1 = 0.5$. It can be shown that $\Delta_1 = 2.28 \times 10^{-2}s$ and $\Delta_2 = 1.14 \times 10^{-2}s$. As a comparison, their values in [106] can be computed as $\Delta_1 = 1.07 \times 10^{-2}s$ and $\Delta_2 = 5.4 \times 10^{-3}s$, respectively. Therefore, in our scheme, the upper bounds of impulsive interval are greatly improved. In the experiment, the time intervals are chosen as $t_{2k} - t_{2k-1} = 2 \times 10^{-2}s$ and $t_{2k+1} - t_{2k} = 1 \times 10^{-2}s$. The initial conditions are given by $[-2.12; -0.05; 0.8]$ and $[-0.2; -0.02; 0.1]$, respectively. That is, the two Chua's circuits are initially not synchronized. The simulation result is given in Fig. 5.6. It is shown that the synchronization errors approach zero very quickly under the impulsive control.   □

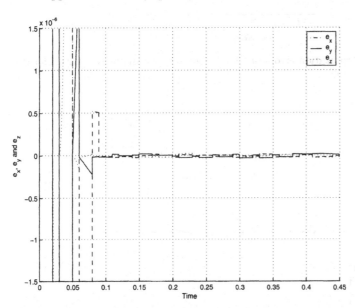

Fig. 5.6. The time response of $e(t)$ of Chua's circuits with impulsive control

Example 5.4.2. In this example, we shall verify the synchronization performance with changes in the values of the design parameter L. With the same system parameters as in Example 5.1.1, if we set $L = 1/300$ and choose $t_{2k} - t_{2k-1} = 2 \times 10^{-2}s$, we have $n_{L=1/300} = 3$ for $\eta = 10^{-6}$. Then $T_{syn} = 0.09s$. If $L = 1/80$, we have $\Delta_1 = 12.74 \times 10^{-2}$, and choose $t_{2k} - t_{2k-1} = 12 \times 10^{-2}s$, we have $n_{L=1/80} = 4$ for $\eta = 10^{-6}$. Then $T_{syn} = 0.72s$. The simulation results are shown in Figs. 5.7 and 5.8. It is illustrated that with a larger L, the bound of impulsive intervals is larger but a longer time is required to synchronize the two systems.   □

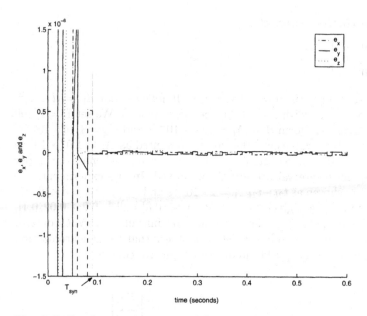

Fig. 5.7. Synchronization errors of Chua's circuits with impulsive control for $\eta = 10^{-6}$ and $L = 1/300$

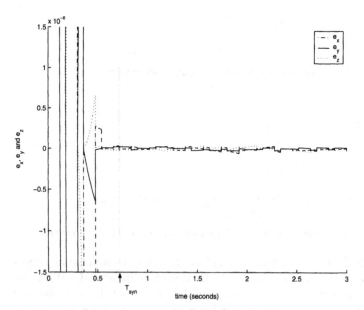

Fig. 5.8. Synchronization errors of Chua's circuits with impulsive control for $\eta = 10^{-6}$ and $L = 1/80$

The approach of linear matrix inequalities are used to further relax the condition on the impulsive synchronization of Chua's circuits as shown in the following theorem [78]:

Theorem 5.4.1. *The error system given by (5.40), is asymptotically stable if*

$$0 \leq a_{min} \leq -\frac{2}{(1+\epsilon_1)\Delta_1} ln(\frac{d_1}{L}), 0 < L < 1, \tag{5.44}$$

where a_{min} denotes the solution of the following optimization problem:

$$min_p\{a\}, \qquad such\ that \qquad (A+H)^T P + P(A+H) \leq aP, \tag{5.45}$$

and d_1 satisfies that

$$(I+B)^T P(I+B) < d_1 P \ and\ d_1 < L, \tag{5.46}$$

where matrix P is symmetric positive definite given by

$$P = \begin{bmatrix} 1 & 0 & 0 \\ 0 & p_1 & p_2 \\ 0 & p_2 & p_3 \end{bmatrix}. \tag{5.47}$$

Proof. Choose the Lyapunov function as $V(e) = e^T Pe$. It can be shown that

$$2e^T P\phi(X, X') \leq e^T (PH + H^T P)e\ ;\ t \neq t_k, \tag{5.48}$$

$$V(e(t_k^+)) \leq d_1 V(e(t_k^-)). \tag{5.49}$$

It follows from (5.40) that

$$\dot{V}(e) = e^T(A^T P + PA)e + \phi^T(X, X')Pe + e^T P\phi(X, X') \tag{5.50}$$
$$= e^T(A^T P + PA)e + 2e^T P\phi(X, X')$$
$$\leq e^T(A^T P + PA)e + e^T(PH + H^T P)e$$
$$= e^T(A+H)^T P + P(A+H))e$$
$$\leq a_{min}V(e).$$

Obviously, the error system given by (5.40) is asymptotically stable.   □

Remark 5.4.2. Based on (5.44), the impulse interval bound for the new scheme can be computed as

$$\Delta_{max} = -\frac{2ln(\frac{d_1}{L})}{(1+\epsilon_1)a_{min}}. \tag{5.51}$$

As $a_{min} < \chi(A, H)$, so the impulsive interval bound with the new scheme is further enlarged.   □

Note that for any given a, inequality (5.45) can be easily solved by the method of linear matrix inequalities. The following algorithm is set up to compute a_{min} [78]:

Algorithm 5.1 *Computation of a_{min}.*

Step 1: *Let $L = 0, U = \chi(A, H)$.*

Step 2: *Let $a_i = \frac{L+U}{2}$.*

Step 3: *Compare the value of a_i with the value of last iteration, set a tolerance η which is a very small number to determine if the iteration should end. If the value of the difference is below the tolerance, go to step 7. Otherwise, if the value of the difference is above the tolerance, go to Step 4. Otherwise, the algorithm is terminated. (For example,we can choose $\bar{\varepsilon} = 10^{-6}$. For the first iteration, a_0 is set to a very small value to let the program run $(a_0 = 10^{-6})$.)*

Step 4: *Check if there is a positive definite matrix P that satisfies the inequality (5.45). If the inequality (5.45) is valid, go to Step 5. Otherwise, go to Step 6.*

Step 5: *Let $U = a_i$, go to Step 2.*

Step 6: *Let $L = a_i$, go to Step 2.*

Step 7: *Obtain the smallest value of a_i.*

5.5 Synchronization of Lorenz Systems via Impulsive Control

In this section, we shall study the impulsive synchronization of two Lorenz systems [203, 198]. The matrices A and $\phi(X)$ are as defined in (1.22) and (1.23), and H is

$$H = \begin{bmatrix} 0 & J & J \\ 0 & 0 & 0 \\ 0 & 0 & 0 \end{bmatrix}, \tag{5.52}$$

where J is the upper bound of both $|y(t)|$ and $|z(t)|$.

Let

$$\Psi(X, \tilde{X}) = \psi(X) - \psi(\tilde{X}) = \begin{bmatrix} 0 \\ -(xz - \tilde{x}\tilde{z}) \\ xy - \tilde{x}\tilde{y} \end{bmatrix}. \tag{5.53}$$

Then, the error system of the impulsive synchronization is

$$\begin{cases} \dot{e} = Ae + \Psi(X, \tilde{X}) \, ; \, t \neq t_k \\ \Delta e|_{t=t_k} = Be \, ; \, k = 1, 2, \cdots \end{cases} \tag{5.54}$$

We can obtain the following result:

Corollary 5.5.1. *The impulsive synchronization of two Lorenz systems, with the error system given by (5.54), is asymptotically stable if*

$$0 \leq \chi(A, H) \leq -\frac{2}{(1 + \epsilon_1)\Delta_1} ln(\frac{d_1}{L}), \tag{5.55}$$

where $0 < L < 1$ and $d_1 = \rho^2(I + B) < 1$, and $d_1 < L$.   □

Remark 5.5.1. Note that the sufficient condition given in [198] is given by

$$0 \leq 2J + \lambda_{max}(A + A^T) \leq -\frac{2}{(1 + \epsilon_1)\Delta_1} ln(\frac{d_1}{L}). \tag{5.56}$$

Clearly, our result is less conservative than that derived in [198]. Meanwhile, the results derived in [198] are less conservative than those obtained in [203]. Therefore, our result is also less conservative than that in [203].   □

Remark 5.5.2. Consider the case that $\epsilon_1 < 1$. For any $0 < L < 1$ satisfying $d_1 \leq L$, which is required by [203], we choose that

$$\Delta_1 = -\frac{2ln(\frac{d_1}{L})}{(1 + \epsilon_1)\chi(A, H)}. \tag{5.57}$$

It can be shown that (5.55) holds. Thus, the origin of (1.27) is asymptotically stable.

Note that Δ_1 obtained in [198] and Δ_{max} obtained in [203] are

$$\Delta_1 = -\frac{2ln(\frac{d_1}{L})}{(1 + \epsilon_1)(2J + \lambda_{max}(A + A^T))}, \tag{5.58}$$

$$\Delta_{max} = -\frac{ln(\frac{d_1}{L})}{2J + \lambda_{max}(A + A^T)}. \tag{5.59}$$

Obviously, a larger bound can be obtained by using our proposed approach.
    □

Two numerical examples are used to illustrate the synchronization of two Lorenz systems.

Example 5.5.1. Suppose that the parameters of the Lorenz systems are: $\sigma = 10$, $\vartheta_3 = 28$, $\vartheta_4 = 8/3$ and the initial conditions are $X(0) = [-36, 30, 0.5]^T$, and $\tilde{X}(0) = [20, -22, 3]$.

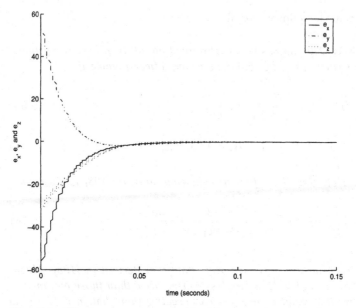

Fig. 5.9. The time responses of $e(t)$ of Lorenz systems with impulsive control

The matrix B is chosen as

$$B = \begin{bmatrix} l & 0 & 0 \\ 0 & -0.1 & 0 \\ 0 & 0 & -0.1 \end{bmatrix},$$

where l is a constant satisfying $-2 < l < 0$. It is easy to show that

$$d_1 = \begin{cases} (l+1)^2, & \text{if } (l+1)^2 > 0.81 \\ 0.81, & \text{otherwise} \end{cases},$$

$$J = 50,$$

$$\lambda_{max}(A + A^T) = 28.051,$$

$$\chi(A, H) = 90.1818.$$

Consider the example with $l = -0.1$, $L = 1/1.05$ and $\epsilon_1 = 0.5$. It can be easily computed that the upper bounds of the impulsive intervals are $\Delta_1 = 2.4 \times 10^{-3}$ and $\Delta_2 = 1.2 \times 10^{-3}$.

The respective values of Δ_1 and Δ_2 in [198], and Δ_{max} in [203] are

$$\Delta_1 = 1.7 \times 10^{-3}, \Delta_2 = 0.85 \times 10^{-3},$$
$$\Delta_{max} = 1.3 \times 10^{-3}.$$

The synchronization errors of the Lorenz systems with the impulsive intervals $t_{2k} - t_{2k-1} = 2 \times 10^{-2}s$ and $t_{2k+1} - t_{2k} = 1 \times 10^{-2}s$ are illustrated in Fig. 5.9, respectively. □

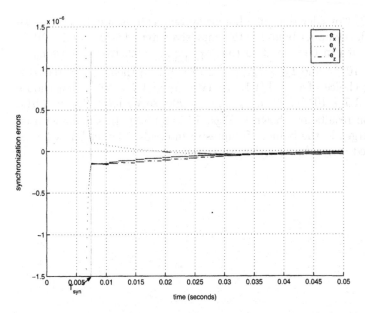

Fig. 5.10. Synchronization errors of Lorenz's system with impulsive control for $\eta = 10^{-6}$ and $L = 1/1.23$

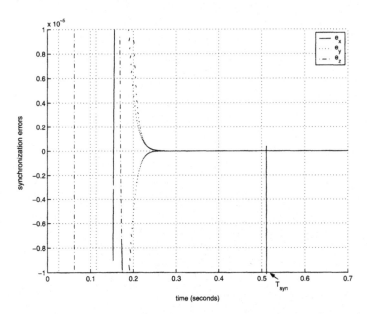

Fig. 5.11. Synchronization errors of Lorenz's system with impulsive control for $\eta = 10^{-6}$ and $L = 1/1.1$

Example 5.5.2. Consider the same Lorenz system as in Example 5.5.1, if we set $L = 1/1.23$, the upper bound of the impulsive interval is $\Delta_1 = 5.5 \times 10^{-5}$. For $\eta = 10^{-6}$, it can be computed that $n_{L=1/1.23} = 92$. If we choose $t_{2k} - t_{2k-1} = 5.5 \times 10^{-5}s$ and $t_{2k} - t_{2k-1} = 2.75 \times 10^{-5}s$, then $T_{syn} = 0.0075s$. With another choice of $L = 1/1.1$, we have $\Delta_1 = 1.7 \times 10^{-3}$, and choose $t_{2k} - t_{2k-1} = 1.7 \times 10^{-3}s$. Then $n_{L=1/1.1} = 200$, so we have $T_{syn} = 0.51s$. The simulation results are shown in Figs. 5.10 and 5.11. It is demonstrated that with a larger L, the bound of impulsive intervals is larger but a longer time is required to synchronize the two systems. □

6. Chaos Based Secure Communication Systems

In this chapter, the results obtained in the previous chapter will be used to design a chaos based secure communication system. Specifically, these results will be used to design a switched and impulsive control strategy for the synchronization of an encrypter and a decrypter. With our switched and impulsive control strategy, the time necessary to synchronize them is minimized while the bound of the impulsive interval after they are synchronized is maximized. With a larger impulsive interval, the transmission efficiency of the chaos based secure communication systems is significantly improved because less bandwidth is needed to transmit the synchronization impulses. Meanwhile, a concept of magnifying-glass and a novel sampling scheme are introduced to improve the security. The proposed system can be used to transmit text, speech, image files, and any digital binary data.

6.1 Secure Communication and Chaotic Systems

Nowadays, the universal availability of information and services enriches our daily lives. However, the exchange of information in many applications such as in commerce, internet banking, security service etc come with significantly risk. The need for secure communication of certain information has thus become more and more important [164, 126, 182].

One of the most popular approaches is to use cryptography to ensure the security of information in communications. Imagine that a sender wants to send a message to a receiver securely such that an eavesdropper cannot read the message. The message is also called the plaintext. The sender usually disguises the plaintext into an apparently random sequence. This is an encryption [164]. The encrypted message is also called the ciphertext. When the receiver has the cipertext, he/she needs to turn the ciphertext back into the plaintext. This is called a decryption [164]. Both the encryption and decryption depend on a key. The key might be any one of a large number of values. There are two general types of key-based algorithms [164]: secret-key or symmetric algorithms, and public-key or asymmetric algorithms. In the public-key cryptography, the sender uses the public key of the receiver to generate the ciphertext, and sends the cipertext to the receiver. The receiver uses his/her private key to recover the message. In the secret-key cryptography, the sender and receiver have previously agreed on the key. The same key is used for both the encryption and decryption.

The key must be kept secret from potential eavesdroppers. The range for the possible values of the key is called the key space [164]. One popular method to attack cryptographic algorithms is an exhaustive key search algorithm, in which an attacker attempts to decrypt a message by systematically trying every key in the key space. Therefore, there exists only one really secure scheme, one-time pad [164], in which the length of the key is the same as that of the plaintext.

Chaotic systems and their applications to secure communications have received a great deal of attention since Pecora and Carroll proposed a method to synchronize two identical chaotic systems [152, 205, 202, 109]. It has been claimed that a chaotic carrier offers possibly several advantages over classical carrier signal like frequency spreading and security can be offered at no extra cost. It is also very competitive when compared with standard communication techniques due to inexpensive implementations [87, 109].

The highly unpredictable and random nature of chaotic signals is the most attractive feature of chaos based secure communication. A chaos-based cryptosystem is essentially a secret-key stream cipher. The parameters of a chaotic system are the key seeds. The running-key sequences are the chaotic signals. The sender and receiver should have the exact key seeds. The encryption running-key sequences are generated by the chaotic system in the sender, while the decryption running-key sequences are produced by an identical chaotic system in the receiver. In these schemes, the chaotic signals are used in the classical cryptographic technique to enhance the degree of security.

The chaos-based secure communication systems have already evolved into the fourth generations [205]. The continuous synchronization is adopted in the first three generations while the impulsive synchronization is used in the fourth generation. The variation in the first three generations lies in the methods for injecting an information at the transmitter and recovering it at the receiver. The first generation relies on the use of additive chaos masking and/or chaotic shift keying. The second generation employs the method of chaotic modulation. Chaotic modulation for analog communications is a more advanced method. It influences the chaos-generating transmitter system by the information-carrying signal via some invertible function. The third generation is generally known as a chaotic cryptosystem. This scheme is proposed for the purpose of improving the degree of security to a much higher level than the first two generations. In this scheme, the combination of the classical cryptographic technique and the chaotic synchronization is used to enhance the degree of security.

The major problem of the continuous synchronization is that these systems are vulnerable under the proposed attacks [170, 154, 171, 218, 141]. Another problem of the continuous chaotic synchronization schemes is the low efficiency of channel usage since the continuously transmitted synchronization signal uses a bandwidth comparable to that of the message signal. Furthermore, today's communications are mostly digital communications. Motivated by these challenges, the idea of impulsive synchronization has been applied in the fourth generation [202, 109]. The impulsive control synchronization method improves the usage of the bandwidth greatly. Less than 94Hz of bandwidth is needed to transmit the synchronization signal for a third-order chaotic transmitter in the impulsive control synchronization while 30 kHz of bandwidth is required for transmitting the synchronization signals in the first three generations [205].

6.2 A Digital Chaos-Based Secure Communication System

6.2.1 System Block Diagram

In this section, we shall present the structure of our proposed digital chaotic secure communication system. The system block diagram is shown in Fig. 6.1. Our scheme is based on a one time pad, which is ideally an unbreakable cipher system [164]. It consists of a nonrepetitive and truly random key of letters or characters, and each key is used exactly once and for only one message. In its original form, it was a one-time tape for teletypewriters [164]. The sender uses each key letter on the pad to encrypt exactly one plaintext character by an addition modulo 26 of the plaintext character and the key character. This kind of cryptograph method produces a completely random ciphertext by the random key sequence added to a non-random plaintext. However, it is very difficult to be widely used in practice, such as in Internet applications. This is because for applications involving considerable amount of information, a large memory is needed to store the random and non-reused key bits. Furthermore, one has to make sure that the sender and receiver are perfectly synchronized.

The chaos based secure communication system is composed of two major parts: the encrypter and decrypter. The input of the cryptosystem is the plaintext, which can be a stream of bits, a text file, a bitmap, a stream of digitized voice, a digital video image, and so on. The original message should be compressed before encryption. There are two reasons why we should use a data compression algorithm together with an encryption algorithm [164]:

1. Cryptanalysis relies on exploiting redundancies in the plaintext and compressing a file before encryption can reduce these redundancies;

2. Encryption is a time-consuming process, compressing a file before encryption speeds up the entire process. Therefore, the original signal is

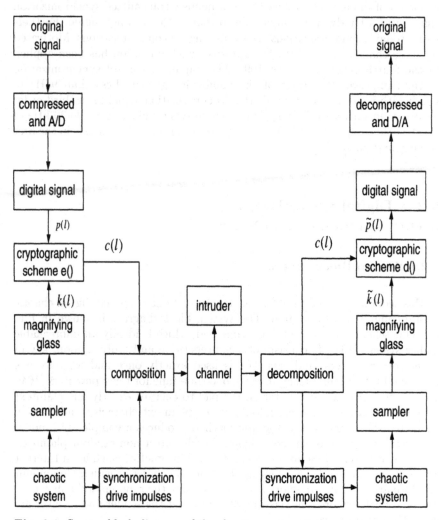

Fig. 6.1. System block diagram of the chaotic cryptosystem

first compressed according to its file type and the plaintext input in the cryptosystem is the compressed digital signal.

Obviously, a chaos-based cryptosystem is essentially a stream cipher system. In the stream cipher cryptography, a critical problem is to efficiently generate a long running-key sequence from a short and random key, which is also called a key seed [164]. The running-key sequence is generated by a function of the key seed – a keystream map F, which in the classical case is a balance boolean function and in the chaotic case is a chaotic dynamical function. The set of all the keystream maps F with a certain input length l

usually constitutes the key space of a stream cipher. For the classical stream ciphers, there exist only a limited number of such functions for a given input length. In contrast, the number of chaotic dynamical functions is theoretically unlimited, since an arbitrary amount of continuous parameters can be included without losing the desired statistical properties.

In a chaos-based cryptosystem, the parameters of the chaos systems play the same role as the random seed and the state variables of the system generated from the chaotic dynamics are the running-key sequences. Since the behavior of chaotic systems are sensitive to initial conditions as well as parameter variations, the parameter space of chaotic systems is large enough for the random key seeds. The running-key sequence can be recovered at the receiver side by synchronizing the chaotic systems. Thus, the chaotic system can be used to generate the one-time-pad-like key sequences in the cryptosystem.

6.2.2 The Encrypter and Decrypter

The encrypter consists of a continuous-time chaotic system, a sampler, a magnifying-glass and a classical encryption function $e(\cdot)$. The encryption algorithm combines the plaintext $p(l)$ and the running-key sequence $s(l)$ to generate the ciphertext $c(l)$. The decrypter is composed of an identical chaotic system, a corresponding sampler, a magnifying-glass, a synchronization controller and a decryption function $d(\cdot)$. The composition block is used to combine the ciphertext and synchronization pulses in a time frame, which is transmitted through the public channel to the receiver. In the receiver, they are separated in the decomposition block. When the chaotic system in the receiver is synchronized with that in the transmitter, the decrypter reproduces the running-key sequence $\tilde{s}(l)$, same as that generated in the encrypter, $s(l)$. Then, the plaintext can be recovered. The details of the encrypter and decrypter are described below.

1. **The Encrypter**

   The chaotic system in the encrypter is given in (5.1) with $r = 3$ and $C = I$. The parameters of the chaotic system are the key seeds and the decrypter needs to have the same parameters to ensure the synchronization. Since the signals are transmitted through a digital channel, the synchronization impulses should be first quantized by a predefined quantizer $Q(\cdot)$, which depends on an amplification factor K used in the magnifying-glass. Since chaos is very sensitive to the initial condition, the quantization error should be less than certain values to ensure that the encrypter and decrypter can be synchronized.

   The state variables of the chaotic system are used to produce the running-key sequences. To further improve the desired randomness in the running-key sequences, the continuous-time chaotic signals are first sampled with a sampling period, T_s, that is considerably below the bandwidth of the chaotic signal. In this way, it will produce frequency aliasing and hence the running-key sequences will have a high degree of randomness for secure encryption. The sampling period acts as one additional key seed in our system. The sampler is defined as follows:

$$X_s(l) = \sum_{m=0}^{+\infty} X(mT_s)\delta(l - mT_s), \tag{6.1}$$

where $X_s(l) = [x_s(l), y_s(l), z_s(l)]^T$ is the vector of the sampled chaotic signals, T_s is the sampling period.

To further enhance the security of the cryptosystem, a magnifying-glass is introduced to increase the parameter sensitivity of the chaotic self-synchronization system. The magnifying-glass is composed of an amplifier and an observer:

The amplifier:

$$s'(l) = K(x_s^2(l) + y_s^2(l) + z_s^2(l))^{1/2}, \tag{6.2}$$

The observer:

$$s(l) = \lfloor s'(l) \rfloor, \tag{6.3}$$

where K is a large number which can be chosen to influence the sensitivity of the system, $\lfloor a \rfloor$ is the integer truncation of a, and $s(l)$ is the running-key sequence. In our system, K is a key design parameter and acts as another key seed.

The encryption algorithm is

$$c = e(p, k), \tag{6.4}$$

where p and c are the plaintext and ciphertext, respectively, $e(\cdot)$ is an applied stream cipher function and can be chosen according to different system demands. The ciphertext is transmitted to the receiver together with the synchronization impulses.

2. **The Decrypter**

The chaotic system with the impulsive controller in the decrypter is

$$\begin{cases} \dot{\tilde{X}}(t) = A\tilde{X}(t) + \phi(\tilde{X}(t)) \; ; \; t \neq t_k \\ \tilde{X}(t_k^+) = \tilde{X}(t_k^-) - B(Q(X(t_k^-)) - \tilde{X}(t_k^-)) \end{cases}, \tag{6.5}$$

where B is the controller matrix to be designed according to the synchronization condition, $Q()$ is the predefined quantizer, t_k is the time instant defined by (5.32) and (5.33).

In the decrypter, the plaintext is recovered via

$$\tilde{s}(l) = \lfloor K(\tilde{x}_s^2(l) + \tilde{y}_s^2(l) + \tilde{z}_s^2(l))^{1/2} \rfloor, \tag{6.6}$$

$$\tilde{p} = d(c, \tilde{s}), \tag{6.7}$$

where $[\tilde{x}_s(l), \tilde{y}_s(l), \tilde{z}_s(l)]$ are the sampled sequences of the chaotic signals in the decrypter, and $d()$ is the corresponding decryption function. The running-key sequence $\tilde{k}$ is produced in the decrypter, and to recover the plaintext it should approximate k.

6.2.3 The Synchronization
of the Encrypter and Decrypter

Corollaries 5.4.2 and 5.5.1 are used to design a switched and impulsive control for the synchronization of the encrypter and decrypter. We have the following two observations from Corollaries 5.4.2 and 5.5.1:

1. With a smaller L, less time is required to synchronize the two chaotic systems while the bound of the impulsive intervals after two systems are synchronized is smaller.

2. With a larger L, more time is required to synchronize the two chaotic systems while the bound of impulsive intervals after the two systems are synchronized is larger.

To obtain a tradeoff between the synchronization time and the bound of the impulsive intervals after the two systems are synchronized, an intermediate value of L is generally chosen in the existing schemes. Better still, the advantage of the two choices can be combined to design a two-stage impulsive control strategy. In the first stage, a small L is chosen to minimize the time necessary to synchronize the two systems. After the two systems are synchronized, a large L is selected to maximize the bound of impulsive intervals. Therefore, a simple switched and impulsive control, which we called it a two-stage impulsive control, can be formulated as follows:

Stage 1. A small L is chosen as L_1, and the impulsive intervals for synchronization can be chosen as $t_{2k} - t_{2k-1} = \Delta_1(L_1)/4$ and $t_{2k+1} - t_{2k} = \Delta_2(L_1)/4$, say.

Stage 2. In this stage, the constant L is selected to be L_2 which is as larger as possible. The upper bounds of each impulsive interval can then be chosen as $\Delta_1(L_2)$ and $\Delta_2(L_2)$, respectively.

To maintain the synchronization of the chaotic systems after the impulsive intervals are increased, it is required to properly define the time instant at which the system is to switch from stage 1 to stage 2. This is achieved by the following results where the effect of quantization is taken into account.

Lemma 6.2.1. *For any $\eta > 0$, define an $n_{0q}(\eta)$ as*

$$n_{0q}(\eta) = \log_{\tilde{L}} \frac{\|e_0\|}{\eta - \frac{\tilde{L}_w}{\tilde{L}-1} q}, \tag{6.8}$$

where e_0 is the initial error vector, q is the quantization parameter, and

$$\tilde{L} = \frac{1}{L_1} e^{\frac{3}{8}\chi(A,H)(\Delta_1(L_1)+\Delta_2(L_1))},$$

$$w = d_1^{1/2} c_1 (e^{\chi(A,H)\frac{\Delta_1(L_1)+\Delta_2(L_1)}{8}} + e^{\chi(A,H)\frac{\max\{\Delta_1(L_1),\Delta_2(L_1)\}}{8}}),$$

$$c_1 = \frac{\|B\|}{2} + q^{-1/2}(\|B\|d_1^{1/2} \sup \|e(t_k^-)\|)^{1/2}.$$

If $k \geq n_{0q}(\eta)$, then $\|e(t_{2k}, t_0, e_0\| < \eta$.   □

Theorem 6.2.1. *For any $\eta > 0$, we denote*

$$\tilde{\eta} = \min\{(\eta - \tilde{w}q)\frac{1}{L_2}, (\frac{\eta}{e^{\chi(A,H)\max\{\Delta_1(L_2),\Delta_2(L_2)\}}} - c_1 q)d_1^{-1/2}\}, \tag{6.9}$$

where

$$\tilde{w} = d_1^{1/2} c_1 (e^{\chi(A,H)\frac{\Delta_1(L_1)+\Delta_2(L_1)}{2}} + e^{\chi(A,H)\frac{\max\{\Delta_1(L_1),\Delta_2(L_1)\}}{2}}). \tag{6.10}$$

When $k \geq n_{0q}(\tilde{\eta})$, we have $\|\|X\| - \|\tilde{X}\|\| < \eta$.   □

With the above theorem, the estimate of the required synchronization time is given by

$$T_{syn} = n_{0q}(T_1 + T_2), \tag{6.11}$$

where $T_1 \leq \Delta_1(L_1)$ and $T_2 \leq \Delta_2(L_1)$ are impulsive time intervals in the first stage.

Two numerical example are used to illustrate the application of two-stage impulsive control strategy.

Example 6.2.1. Choose the parameters of Chua's circuit as $\vartheta_1 = -0.722451121$, $\vartheta_2 = -1.138411196$, and the matrix A as

$$A = \begin{bmatrix} -9.35159085 & 9.35159085 & 0 \\ 1 & -1 & 1 \\ 0 & -14.790313805 & 0.016073965 \end{bmatrix}. \tag{6.12}$$

At the first stage, L is chosen as $1/300$. It can be easily computed that

$$\Delta_1(1/300) = 2.275 \times 10^{-2} \ ; \ \Delta_2(1/300) = 1.137 \times 10^{-2}.$$

After the two systems are synchronized, L is chosen as $1/80$. It follows that

$$\Delta_1(1/80) = 12.74 \times 10^{-2} \ ; \ \Delta_2(1/80) = 6.37 \times 10^{-2}.$$

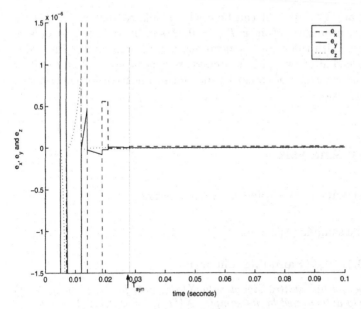

Fig. 6.2. Synchronization errors of Chua's circuits with the switched and impulsive control strategy for $\eta = 10^{-6}$

The impulsive intervals are set as $T_1 = 0.005s$ and $T_2 = 0.002s$ respectively in the first stage and after the two Chua's circuits are synchronized, they are set as 0.12s and 0.06s respectively in the second stage. The synchronization error with the two-stage impulsive control synchronization strategy is shown in Fig. 6.2. Compared the results with Example 5.4.1, it can be illustrated that synchronization can be achieved rapidly and the synchronization is maintained even though the impulsive intervals used are larger. As a result, it will improve the efficiency of channel bandwidth utilization when applied to chaos-based secure communications.   □

Example 6.2.2. We have already examined the synchronization performance with respect to changes in the design parameter L in Example 5.4.2. Here, we will verify the performance of the switched and impulsive control strategy. The parameters of the two Chua's circuits and the impulsive controller are the same as in Example 5.4.2.

Note that $\chi(A, H) = 16.8385$ and $d_1 = 0.0025$. We choose $L_1 = 1/300$ at the first stage. After the two Chua's circuits are synchronized, we choose $L_2 = 1/80$. Then we have $\Delta_1(L_1) = 2.275 \times 10^{-2}$, $\Delta_2(L_1) = 1.137 \times 10^{-2}$, $\Delta_1(L_2) = 12.74 \times 10^{-2}$ and $\Delta_2(L_2) = 6.37 \times 10^{-2}$.

The impulsive intervals are set as $T_1 = 5 \times 10^{-3}s$ and $T_2 = 2 \times 10^{-3}s$ in the first stage and after the two Chua's circuits are synchronized, they are set as $T_1 = 0.12s$ and $T_2 = 0.06s$ in the second stage. The quantizer step is

$q = 10^{-5}$. For any $\eta = 10^{-6}$, it can be easily calculated that $n_{0q} = 4$, thus the required synchronization time is $T_{syn} = 0.028s$ in the two-stage impulsive control strategy. With the 2-stage synchronization techniques, the impulsive interval is enlarged after two Chua's circuits are synchronized. As a result, it will improve the efficiency of channel bandwidth utilization for chaos-based secure communications.   □

6.3 Security Analysis

6.3.1 The Security of Sampled Chaotic Signals

Let us recall the sampling theorem:

Theorem 6.3.1. *[146]* **Sampling Theorem**

Let $X(t)$ be a band-limited signal with $\hat{X}(jw) = 0$ for $|w| > w_M$. Then $X(t)$ is uniquely determined by its samples $X(lT_s)$, $l = 0, \pm1, \pm2, \cdots$ if

$$w_s > 2w_M, \tag{6.13}$$

where

$$w_s = \frac{2\pi}{T_s}. \tag{6.14}$$

Given these samples, $X(t)$ can be reconstructed by generating a periodic impulse train in which successive impulses have amplitudes that are the sampled values. This impulse train is then processed through an idea lowpass filter with gain T and cutoff frequency greater than w_M and less than $w_s - w_M$. The resulting output signal will exactly equal $X(t)$.   □

However in our application, we do not want the sample X_s to be recovered by an intruder. We want X_s to exhibit a white noise liked characteristics.

It is well known that the white noise is characterized by its uniformly distributed frequency spectrum over the whole frequency band. In the time domain, its correlation function is an impulsive response. From the *Sampling Theorem*, we know that when a chaotic signal is sampled with a frequency below its bandwidth, the frequency spectrum of its sampled discrete time sequence can be flatten due to frequency aliasing. In this sense, the sampled chaotic sequence is whiten and its randomness is improved.

The Lorenz system is now used to illustrate the effect of sampling frequency in the chaos-based cryptosystem.

Example 6.3.1. The parameters of the Lorenz system are $\sigma = 10$, $\vartheta_3 = 28$ and $\vartheta_4 = 8/3$. The initial state is $[-0.2, 1, 10]$. The continuous-time Lorenz system is numerically integrated by the fourth-order Runge-Kutta method with a fixed step size of 10^{-5}.

The correlation function is adopted to quantitatively evaluate the randomness of the running-key sequence and its sensitivity to variation of the sampling frequency. For two discrete-time sequences X and Y of length N, their cross-correlation is

$$R_{XY}(m) = \begin{cases} \sum_{l=0}^{N-m-1} X(l+m)Y(l); & m \geq 0 \\ R_{XY}(-m); & m < 0 \end{cases}. \qquad (6.15)$$

It becomes the autocorrelation R_{XX} of the sequence $X(l)$ if $X(l) = Y(l)$.

To demonstrate the improvement in the randomness of sampled chaotic signals by frequency domain aliasing, the continuous-time chaotic signal is sampled with different sampling periods, $T_s = 0.008t_u$, $0.08t_u$, $0.4t_u$ and $0.8t_u$, respectively. t_u is only the simulation time unit, which need not be the same as the physical time. The experimental results are demonstrated in Fig. 6.3. As expected, when the sampling frequency becomes lower, the frequency spectrum of the sampled signal becomes flatter, which implies that the sampled signal becomes more white-noise like. This effect is due to aliasing in the frequency domain. Further evaluation of the randomness of the sampled signal can be carried out using the autocorrelation function. The autocorrelations of the continuous chaotic signal X and its sampled signal X_s with sampled period $T_s = 0.4t_u$ for 25,000 samples are computed. The resultant autocorrelations of the continuous-time X and that of the sampled X_s are plotted in Figs. 6.4 and 6.5, respectively. It is shown that the autocorrela-

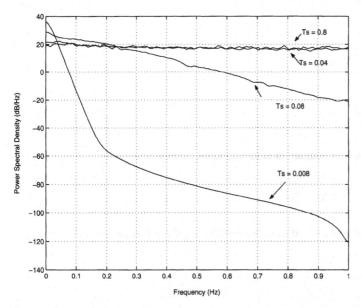

Fig. 6.3. Frequency spectra of X_s with different sampling periods

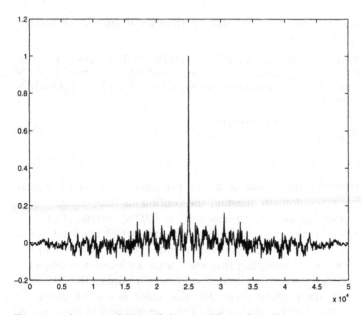

Fig. 6.4. Autocorrelation of the variable x of the Lorenz system

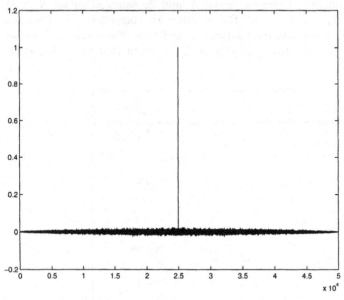

Fig. 6.5. Autocorrelation of sampled variable x_s of the Lorenz system ($T_s = 0.4t_u$)

tion of the sampled signal is more impulse like. This is due to aliasing in the frequency domain and it improves the degree of randomness of the sampled signal.   □

The sampling frequency of chaotic signals should be below its bandwidth such that the running key sequence has a high degree of randomness. There are two possible methods to achieve this. One is to use a fixed frequency sampling scheme with the frequency always below its bandwidth as discussed above. The other is to use a time varying frequency sampling scheme, the overall frequency is not below the bandwidth of the chaotic signal but the frequency is sometimes less than the bandwidth of the chaotic signal. The advantage of the latter one is that it can be used to obtain a good trade-off between security and bandwidth utilization. An example is given below.

The initial sampling frequency is

$$w_s = \gamma w_M, \tag{6.16}$$

where $\gamma (4 > \gamma > 3)$ is a parameter chosen by the user.

Let

$$T_s = \frac{2\pi}{w_s}. \tag{6.17}$$

The sampling function is

$$p(t) = \sum_{i=-\infty}^{+\infty} \delta(t - iT_s). \tag{6.18}$$

The sampled values are denoted by $X(iT_s)$ and are filtered as

$$Y(2j) = X(3jT_s), \tag{6.19}$$
$$Y(2j + 1) = X((3j + 1)T_s). \tag{6.20}$$

where $Y(l)$ is the final sampled value.

The relationship between $Y(l)$ and $X(lT_s)$ can also be represented by

$$Y(l) = \frac{1}{3}\left(\sum_{k=0}^{2} e^{j*2\pi(l-1)k/3} + \sum_{k=0}^{2} e^{j*2\pi(l+1)k/3}\right) X(lT_s)$$
$$= \frac{2 - e^{j2\pi l/3} - e^{j4\pi l/3}}{3} X(lT_s). \tag{6.21}$$

The relationship between their z-transforms is

$$Y(z) = \frac{2}{3}X(z) - \frac{1}{3}X(ze^{j2\pi/3}) - \frac{1}{3}X(ze^{j4\pi/3}). \tag{6.22}$$

Obviously, the overall sampling frequency is

$$\bar{w}_s = \frac{2}{3}w_s > 2w_M. \tag{6.23}$$

However, the resulting output signal will not be exactly equal to $X(t)$. The time varying sampling frequency is a simple switched control technique.

Example 6.3.2. The parameters of the Lorenz system are $\sigma = 10$, $\vartheta_3 = 28$ and $\vartheta_4 = 8/3$. The initial state is $[-36, 30, 0.5]$. The continuous-time Lorenz system is numerically integrated by the fourth-order Runge-Kutta method with a fixed step size of 10^{-4}.

The experimental results are illustrated in Figs. 6.6 and 6.7. The fixed sampling periods are $0.02t_u$, $0.04t_u$, $0.06t_u$ and $0.08t_u$. The time varying periods are $(0.02t_u, 0.06t_u)$ and $(0.03t_u, 0.06t_u)$. The sampled values for $(0.02t_u, 0.06t_u)$ are generated by the following two steps:

Step 1. Generate the initial sampled sequence by $0.02t_u$ and denote them by $X(l)$ $(1 \le l \le +\infty)$.

Step 2. Sample them to generate the desired values by

$$Y(2j + 1) = X((4j + 1)T_s), \tag{6.24}$$
$$Y(2j + 2) = X((4j + 2)T_s). \tag{6.25}$$

Similarly, the sampled values for $(0.03t_u, 0.06t_u)$ are generated with the initial sampled sequence by sampling at $0.03t_u$ and denote them by $X(l)$ $(1 \le l \le +\infty)$. Then sample them to generate the desired values by equations (6.19) and 6.20).

It is clear that the time varying sampling scheme can be used to generate a much flatter chaotic signal than a fixed sampling scheme. □

We shall now provide an example to show that the time varying sampling scheme is also applicable to video coding.

Example 6.3.3. **A Video Coding System** In existing video coding systems, a video sequence is captured by a video camera at a given frame rate, and is then compressed by an encoder according to a given bit rate. There is no feedback information from the encoder to the camera, and the video camera and the encoder are not studied systematically.

Since human eyes are more sensitive to a video segment with high motion, a high sampling frame rate is used for the segment with high motion, while a low sampling frame rate is adopted for that with low motion. In other words, the sampling rate is time varying, and is determined by the encoder according to video content. The video sequence is captured by the video camera according to the frame rate that is fedback from the encoder. The video source is then compressed when it is generated. This is illustrated in Fig. 6.8. □

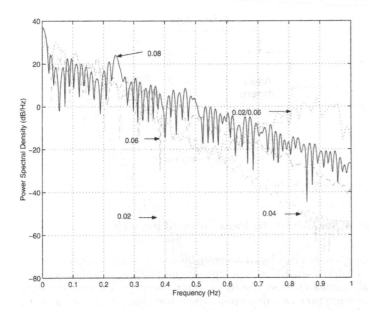

Fig. 6.6. Frequency spectra of X_s with fixed sampling periods and time varying sampling $(0.02t_u, 0.06t_u)$

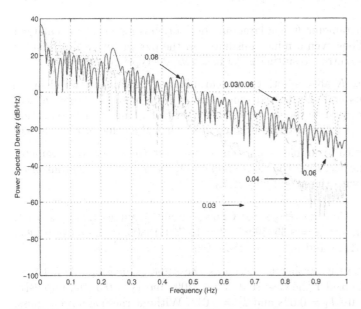

Fig. 6.7. Frequency spectra of X_s with fixed sampling periods and time varying sampling $(0.03t_u, 0.06t_u)$

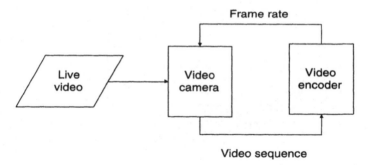

Fig. 6.8. A video coding system

6.3.2 Enhanced Sensitivity to Parameter Mismatch

The magnifying-glass is introduced to transform the sampled chaotic state variables into a keystream before encrypting the plaintext. Assuming that there is a small mismatch that results in errors between the chaotic state variables of the encrypter and decrypter which are given as σ_x, σ_y and σ_z. Then the signal getting through the amplifier becomes

$$\hat{k}(l) = \lfloor K((x+\sigma_x)^2 + (y+\sigma_y)^2 + (z+\sigma_z)^2)^{1/2} \rfloor. \tag{6.26}$$

Since the parameter K is a large number, any mismatch will be enlarged many times. Thus even a minor mismatch in the parameters will produce a large decryption error, resulting in an incorrect decryption key sequence.

The value of K affects both the synchronization time of the two identical chaotic systems and the desired sensitivity of the system. A larger K will produce a more secure system but at the expense of a longer synchronization time. Thus, in practice, a tradeoff is required when the value of K is chosen.

Two numerical examples are given to illustrate the effects of parameter mismatch in the decrypter chaotic system and to examine the synchronization time required with the magnifying-glass.

Example 6.3.4. Consider the same Chua's circuit given in Example 5.1.1. The parameters are $\alpha = 9.35159085$, $\beta = 14.790313805$, $\gamma = -0.016073965$, $\vartheta_1 = -0.722451121$ and $\vartheta_2 = -1.138411196$.

The impulsive controller is chosen as the same as in Example 5.4.1, so $d_1 = 0.0025$. Consider the case that L = 1/300, and $\epsilon_1 = 0.5$, the impulsive time intervals are $T_1 = 0.02s$ and $T_2 = 0.01s$. Without the magnifying-glass, we examine the performance of synchronization when the parameters of the chaotic system α and b in the decrypter have 10% mismatch, respectively. It can be shown from Figs. 6.9 and 6.10 that the synchronization errors are small. Therefore, the synchronization of the chaotic systems is not sensitive enough to the parameter mismatch.

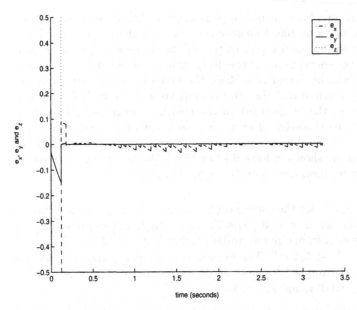

Fig. 6.9. Synchronization errors of Chua's circuits without the magnifying-glass but with 10% mismatch in α

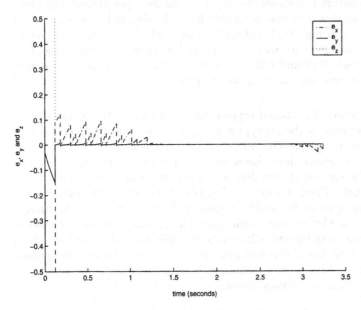

Fig. 6.10. Synchronization errors of Chua's circuits without the magnifying-glass but with 10% mismatch in b

Let us use the amplifier defined in (6.2) and the observer defined in (6.3). The value of K is set as 100. Consider the case that there is just a 1% parameter mismatch in α and a parameters of the decrypter. The keystream errors between the encryption and the decryption are shown in Figs. 6.11 and 6.12. Clearly, with the magnifying-glass, the system is highly sensitivity to the parameter mismatch and the error signals are not stable. It is therefore difficult to recover the original information with approximate parameters. The security of the chaos-based cryptosystem is thus improved. The experimental results on the synchronization time required with and without the magnifying-glass are shown in Figs. 6.13 and 6.14. We can see that it requires a longer synchronization time with the magnifying-glass.   □

Example 6.3.5. Consider the same Lorenz system as that given in Example 5.5.1. The parameters are $\sigma = 10$, $\vartheta_3 = 28$, $\vartheta_4 = 8/3$. In this example, the two Lorenz systems start from different initial states: $X(0) = [-2.12, -0.05, 2.8]^T$ and $\tilde{X}(0) = [-0.2, -0.5, 2.6]^T$. The impulsive controller is chosen to be the same as that in Example 5.5.1 and we set the impulsive intervals as $T_1 = 0.002s$ and $T_2 = 0.001s$, and K $= 100$.

When the parameter r in the decrypter has 1% mismatch, the keystream errors between the encryption and the decryption are shown Fig. 6.15, while the keystream errors are shown in Fig. 6.16 without the magnifying-glass. When the parameter b in the decrypter has 1% mismatch, the simulation results are shown in Figs. 6.17 and 6.18, respectively. The simulation results on the synchronization time required with and without the magnifying-glass are shown in Figs. 6.19 and 6.20, respectively. Clearly, it requires a longer synchronization time with the magnifying-glass.   □

In our proposed chaos-based cryptosystem, to recover the plaintext, the two chaotic systems in the encrypter and decrypter must be synchronized to get the same key sequences. Thus, an intruder who wants to eavesdrop the transmission message must know not only the exact parameters and the structure of the chaotic system but also the synchronization impulses and the sampling rate. Furthermore, the lengths of impulsive intervals are not constant in our system; it would be more difficult to perform the inverse prediction and to identify the synchronization impulses and the scrambled signal if the lengths of impulsive intervals are unknown. All these features in our proposed chaos-based cryptosystem are the results of combining simple switched and impulsive control techniques to greatly enhance the security of the system for secure communications.

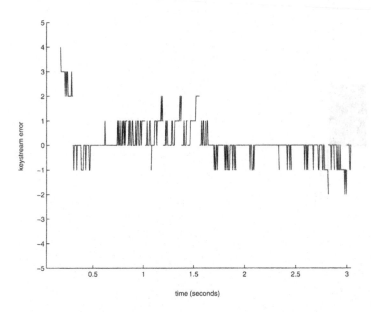

Fig. 6.11. Synchronization errors of Chua's circuits with the magnifying-glass and a 1% mismatch in α

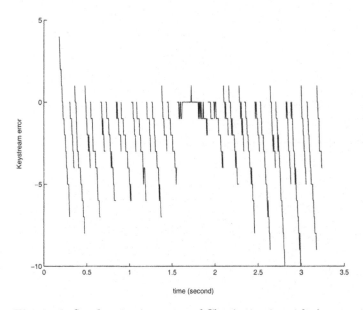

Fig. 6.12. Synchronization errors of Chua's circuits with the magnifying-glass and a 1% mismatch in a

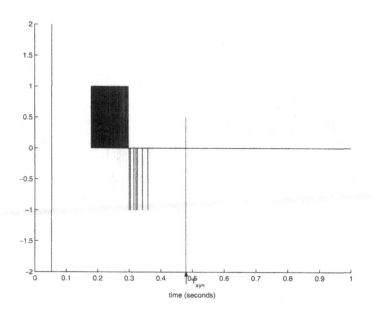

Fig. 6.13. Keystream errors between the encryption and the decryption with the magnifying-glass $K = 100$

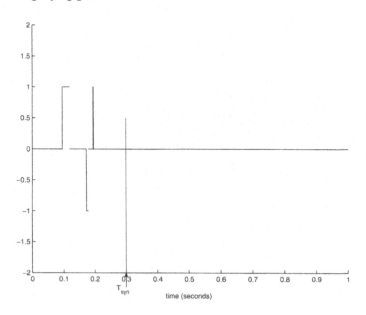

Fig. 6.14. Keystream errors between the encryption and the decryption without the magnifying-glass

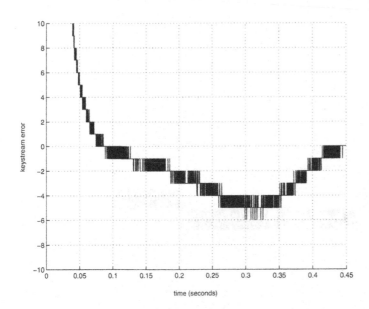

Fig. 6.15. Keystream errors between the encryption and the decryption with 1% mismatch in r and the magnifying-glass K=100

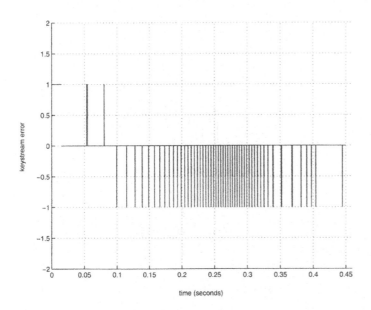

Fig. 6.16. Keystream errors between the encryption and the decryption with 1% mismatch in r but no magnifying-glass

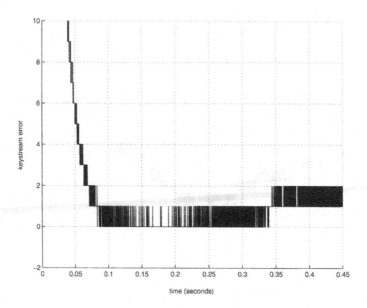

Fig. 6.17. Keystream errors between the encryption and the decryption with 1% mismatch in b and the magnifying-glass K=100

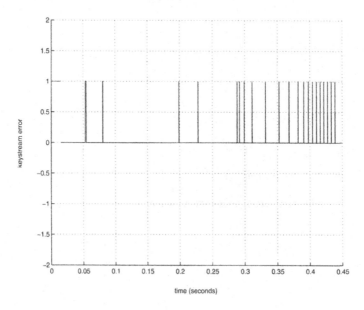

Fig. 6.18. Keystream errors between the encryption and the decryption with 1% mismatch in b but no magnifying-glass

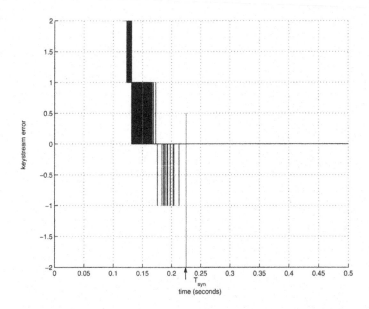

Fig. 6.19. Keystream errors between the encryption and the decryption with the magnifying-glass $K = 100$

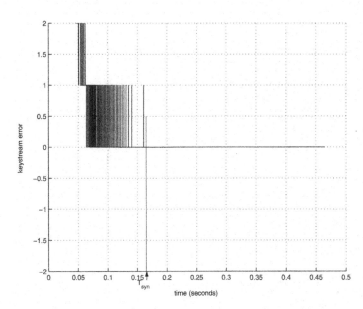

Fig. 6.20. Keystream errors between the encryption and the decryption without the magnifying-glass

7. Scheduling of Switched Server Systems

In this chapter, we shall present an interesting and practical application of the stability results obtained in Chapter 2. Specifically, the stability results are applied to study the scheduling problem of a class of switched server systems which is composed of one server and multiple clients. A simple switching law of the server, *feedback cyclic control policy*, is proposed for the client/server systems. This class of client/server systems can be used to represent many practical systems, such as, a Token/Bus network, a Token/Ring network, a crossroad scheduling system, single-machine flexible manufacturing systems [155], and so on.

In this chapter, we study a simple case, i.e. the arrival rate of each client is constant and it is known in priori. The *feedback cyclic control policy* will be extended in the next chapter to the case that the arrive rate is time varying and cannot be known in advance.

7.1 A Model for Switched Server Systems

We shall present a model for a class of switched and server systems that have one server and several clients.

Consider a switched server system consisting of $n(n \in \{2, 3, \cdots \})$ clients and one server. The work arrives at the ith client continuously at a constant rate $p_i > 0$ and the server removes the work from a client at the unit rate. Obviously, a necessary condition for the existence of a control policy to asymptotically stabilize such a switched server system is that $\sum_{i=1}^n p_i \leq 1$. To simplify the analysis, we study the following two cases separately:

Case 1. $\sum_{i=1}^n p_i = 1$. In this case, the server switches from one client to another client without any reset time.

Case 2. $\sum_{i=1}^n p_i < 1$. In this case, the server should have a positive reset time when it switches from one client to another client.

Let $X_i(t)$ denote the the amount of work of client i at time t. Then, the whole system in case 1 is described by the following n CVSs:

$$\begin{cases} \dot{X}_1(t) = p_1 \\ \cdots \\ \dot{X}_{i-1}(t) = p_{i-1} \\ \dot{X}_i(t) = p_i - 1 \; ; \; i = 1, 2, \cdots, n \; . \\ \dot{X}_{i+1}(t) = p_{i+1} \\ \cdots \\ \dot{X}_n(t) = p_n \end{cases} \tag{7.1}$$

CVS $i(i = 1, 2, \cdots, n)$ corresponds to the process that the server removes the work from client i.

For case 2, the whole system is described by the above n CVSs (7.1) and the following CVS:

$$\begin{cases} \dot{X}_1(t) = p_1 \\ \cdots \\ \dot{X}_i(t) = p_i \\ \cdots \\ \dot{X}_n(t) = p_n \end{cases} \tag{7.2}$$

CVS (7.2) corresponds to the process that the server switches from one client to another one.

7.2 Performance Indices

In this section, we shall describe several desirable performance indices which should be considered when a real time scheduling algorithm is designed for the client/server systems mentioned in the above section.

1. **Stability**: The overall system should be stable with respect to a limit cycle under the designed algorithm. This is a fundamental requirement for a real time scheduling algorithm.

2. **Simplicity**: The designed algorithm should not require too much computation. If the algorithm is too complex, the computation may not be completed in time. Then, the efficiency of the scheduling algorithm will be affected.

3. **Transient performances**: The transient period, which is the time interval for the system to approach the limit cycle, should be as short as possible, and there is no client which occupies the server for an unbearable long time interval.

4. **Steady state performances**: After the transition is over, the maximum buffer level of each client should be as low as possible such that the total awaiting time can be minimized, and the number of switchings of the

server should be as few as possible such that the utilization efficiency of the server is maximized.

5. **Idle time (excluding the set-up time)**: To improve the utilization efficiency of the system, the idle time of the algorithm should be as small as possible during both the transient period and the steady state period.

6. **Fairness**: To ensure the fairness of the system with respect to the arrival rate of each client, the upper bound of the service time should be increased with an increase in the arrival rate. After the transition is over, the service time of each client should also be increased with an increase in the arrival rate.

7.3 Simple Cyclic Control Policies

Since the service rate is fixed at the unit rate, i.e. $K_i(X(t)) = 1$, we only need to determine the switching law of the sub-controllers. The stability results that are obtained in chapter 2 will be used to derive a simple scheduling method, named as a *simple cyclic control policy*, for the switched server systems (7.1) and (7.2).

The server switches instantaneously from one client to another one according to the following policies:

Case 1. **Simple Cyclic Control Policy 1.**

1. The server starts with the first client.

2. As soon as the work of the ith $(i = 1, 2, \cdots, n-1)$ client is finished, the server switches to the $(i + 1)$th one.

3. When the server finishes the work of the nth client, it switches back to the first one.

The policy is generalized from that in Example 1.2.1 and was analyzed by [122] in the case that $X(t_0) \in \{X | \sum_{i=1}^{n} X_i = 1\}$ (where X_i is the work of the ith client). It was shown that the switched server system under the *simple cyclic control policy* has a unique limit cycle and the system is locally asymptotically stable with respect to such a cycle. The result is presented as [122]

Theorem 7.3.1. *Consider a switched server system with $p_i (p_i > 0, \sum_{i=1}^{n} p_i = 1)$ and the simple cyclic control policy 1. There exists an eventual periodic trajectory with the initial condition $X(0)$ in the set defined by Equation (2.20) such that any trajectory with the initial condition $\hat{X}(0)$ in the set converges to the eventual periodic trajectory.* □

Case 2. **Simple Cyclic Control Policy 2.**

1. The server starts with the first client.

2. After the work of the ith ($i = 1, 2, \cdots, n-1$) client is finished, the server switches to the $(i+1)$th one with a positive reset time.

3. When the server finishes the work of the nth client, it switches back to the first one with a positive reset time.

The reset time is defined according to a specific practical situation. Assume that the reset time from the ith buffer to the $(i+1)$th buffer is $\delta_{i,i+1}(i = 1, 2, \cdots, n)$ satisfying $\sum_{i=1}^{n} \delta_{i,i+1} > 0$. It can be easily shown that such a switched server system is globally asymptotically stable with respect to a periodic trajectory which is an invariant set. The result is as follows [122]

Theorem 7.3.2. *Consider the switched server system with p_i ($p_i > 0$, $\sum_{i=1}^{n} p_i < 1$), $\sum_{i=1}^{n} \delta_{i,i+1} > 0$ and the simple cyclic control policy 2. It is globally stable and the eventual period is*

$$T = \frac{\sum_{i=1}^{n} \delta_{i,i+1}}{1 - \sum_{i=1}^{n} p_i}. \tag{7.3}$$

□

Remark 7.3.1. The simple cyclic control policy was proposed by us in [94] while Savkin and Matveev [123, 161, 122, 162] studied the stability of simple cyclic control policy 2 and further used it to study the scheduling problem of a wider class switched flow networks. □

The *simple cyclic control policy* is very simple and the switched server system is globally and asymptotically stable with respect to a limit cycle under this algorithm [122]. Moreover, the stable performance indices are also very nice in the sense that both the switching time of the server and the maximum stable buffer levels are minimized.

However, there may be some clients which will occupy the server for an unbearable long time interval during the transient process. An example is given below.

Example 7.3.1. Consider a switched server system composed of three clients and one server, and the parameters are

$$p_1 = 0.1 \; ; \; p_2 = 0.2 \; ; \; p_3 = 0.6 \; ;$$
$$\delta_{12} = 0.1 \; ; \; \delta_{23} = 0.05 \; ; \; \delta_{31} = 0.06.$$

We consider two sets of initial values and the experimental results are given in Tables 7.1 and 7.2 for the following two cases, respectively.

Table 7.1. The interval during the transient process

service times (k)	client 1	client 2	client 3
1	5.56	5.79	30.74
2	4.08	8.76	19.58
3	3.17	5.74	13.68
4	2.18	4.02	9.61
5	1.54	2.84	6.88
6	1.1	2.04	5.04

Table 7.2. The interval during the transient process

service times (k)	client 1	client 2	client 3
1	11.11	3.93	26.28
2	3.38	7.46	16.58
3	2.7	4.87	11.67
4	1.86	3.44	8.26
5	1.32	2.45	5.97
6	0.96	1.79	4.47

1. $a = 5.0$, $b = 3.5$ and $c = 5.4$.

2. $a = 10.0$, $b = 0.9$ and $c = 1.4$.

Obviously, client 3 occupies the server for an unbearable long time interval during the transient process when the *simple cyclic control policy* is used. □

This is not desirable for a real time scheduling algorithm. Thus, the algorithm should be enhanced by imposing a well defined upper bound on the serving time for each client to avoid the problem. At the same time, the other good performances of the original algorithm should still be maintained.

7.4 Feedback Cyclic Control Policy

We shall provide a method to enhance the *simple cyclic control policy* by imposing an upper bound on the servicing time for each client and use it to study the scheduling problem of the client/server systems. Our algorithm achieves a nice balance among the six performance indices given in Sect. 7.2.

To ensure the stability of the system, the upper bound should be well defined. For simplicity, two notations are defined as

$$\delta = \sum_{l=1}^{n-1} \delta_{l,l+1} + \delta_{n,1},$$

$$b_i = \frac{p_i}{1 - p_i}.$$

Then, the upper bound is

$$T_j = \Gamma * p_j + X_j, \tag{7.4}$$

where $\Gamma > 0$ is an adjustable parameter and it determines the convergence speed based on our simulation studies, $X_j(j = 1, 2, \cdots, n)$ are the time for the server to complete the work of client j during the steady state period and they are the solutions to the following n equations:

$$\left(\sum_{1 \le j \le n, j \ne i} X_j + \delta \right) \times b_i = X_i \; ; \; i = 1, 2, \cdots, n. \tag{7.5}$$

It can be easily computed that

$$X_i = \frac{p_i \delta}{1 - \sum_{j=1}^{n} p_j} \; ; \; i = 1, 2, \cdots, n. \tag{7.6}$$

It can be shown from (7.4) and (7.6) that

$$\frac{T_1}{p_1} = \frac{T_2}{p_2} = \cdots = \frac{T_n}{p_n} = \frac{\delta}{1 - \sum_{j=1}^{n} p_j} + \Gamma, \tag{7.7}$$

$$\frac{X_1}{p_1} = \frac{X_2}{p_2} = \cdots = \frac{X_n}{p_n} = \frac{\delta}{1 - \sum_{j=1}^{n} p_j}. \tag{7.8}$$

Clearly, the upper bounds $T_j(j = 1, 2, \cdots, n)$ are optimal with respect to the following performance index:

$$\min\left\{ \frac{1}{n} \sum_{i=1}^{n} \frac{T_i^2}{p_i^2} - \left(\frac{1}{n} \sum_{i=1}^{n} \frac{T_i}{p_i} \right)^2 \right\}, \tag{7.9}$$

and the steady state service time $X_j(j = 1, 2, \cdots, n)$ are also optimal with respect to the following performance index:

$$\min\left\{ \frac{1}{n} \sum_{i=1}^{n} \frac{X_i^2}{p_i^2} - \left(\frac{1}{n} \sum_{i=1}^{n} \frac{X_i}{p_i} \right)^2 \right\}. \tag{7.10}$$

Equations (7.7)–(7.10) imply that the upper bounds and the steady state service time are fair for all clients with respect to their arriving rates.

Certainly, other types of upper bounds can still work. For example, $T_j = \Lambda_j + X_j$, where $\Lambda_j > 0(j = 1, 2, \cdots, n)$ are adjustable parameters.

After computing the upper bound for the service time of each client, the service time for each client should also be calculated. Suppose that the level of client j is $\tilde{X}_j(k)$ when the server switches to client j at the kth time. Then, the service time for client j at the kth time is

$$S_j(k) = \begin{cases} T_j; & \text{If } \frac{\tilde{X}_j(k)}{1-p_j} > T_j; \\ \frac{\tilde{X}_j(k)}{1-p_j}; & \text{Otherwise} \end{cases}. \tag{7.11}$$

In other words, if the level of client j is too high, then the server switches to other clients after serving the client for T_j units. Otherwise, the server switches to other clients after the work of client j is completed.

With the above service time, our new scheduling algorithm, named **feedback cyclic control policy**, is presented as follows:

1. The server starts with client 1.

2. The server switches from client j to client $(j+1)$ at the kth times when the service time of client j at the kth times is $S_j(k)$ for $j = 1, 2, \cdots, n-1$.

3. The server switches back to client 1 at the $(k+1)$ times when the service time of client n is $S_n(k)$ at the kth times.

We now illustrate the performance of the *feedback cyclic control policy* via a numerical example.

Example 7.4.1. Consider the switched server system given in Example 7.3.1. It can be easily computed that

$$X_1 = 0.21 \ ; \ X_2 = 0.42 \ ; \ X_3 = 1.26.$$

We first show that the system is globally asymptotically stable with respect to a periodic trajectory obtained via the *simple cyclic control policy*. The experimental results are given in Figs 7.1-7.4.

We also consider the relationship between the transient interval and the value of Γ. The experimental results are given in Table 7.3 where the steps are the minimum service times required by each client such that the buffer levels are in the limit cycle. The values within [   ] are the initial values of all clients.

It is clear that the convergence speed is a strictly increasing function of the value of Γ. $\qquad\qquad\square$

Table 7.3. The relationship between the transient interval and the value of Γ

Γ	Steps $[0.9, 0.5, 1.4]^T$	Steps $[10.0, 5.5, 2.4]^T$
0.1	324	2744
0.5	87	571
0.8	66	369
1.0	59	301
2.0	46	167
3.0	42	122

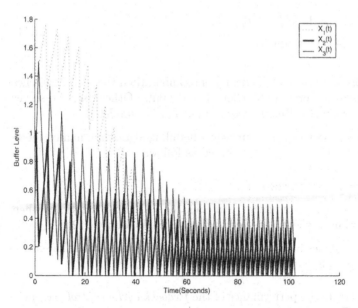

Fig. 7.1. The stability of FCCP with initial states $[1.8, 0.9, 0.5]^T$

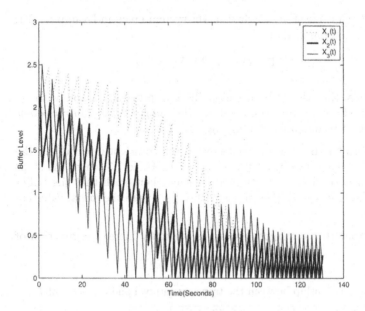

Fig. 7.2. The stability of FCCP with initial states $[2.5, 2, 1.5]^T$

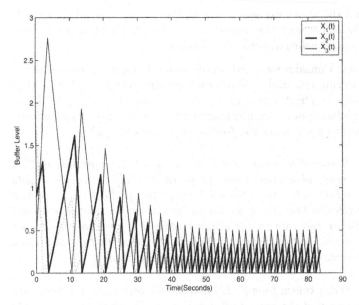

Fig. 7.3. The stability of the *simple cyclic control policy* with initial states $[1.8, 0.9, 0.5]^T$

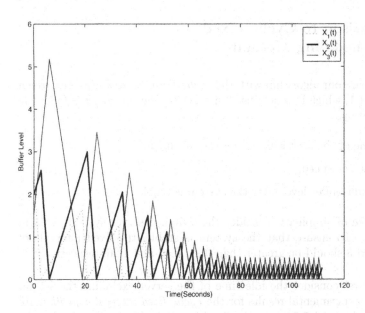

Fig. 7.4. The stability of the *simple cyclic control policy* with initial states $[2.5, 2, 1.5]^T$

Clearly from the above examples, the *feedback cyclic control policy* may also be globally stable with the respect to the limit cycle obtained via the *simple cyclic control policy* as conjectured below:

Conjecture : Consider switched server systems composed of one server with the unit serving rate and n clients with arriving rate p_j, and the set-up time from client i to client j is $\delta_{ij}(> 0)$. Assume that $1 > \sum_{j=1}^{n} p_j$. Then, the system is globally and asymptotically stable with respect to a periodic trajectory (a limit cycle) under the *feedback cyclic control policy*.   □

Remark 7.4.1. It should be mentioned that an important scheduling method named deficit round robin (DRR) was proposed in [174] to provide quality of service in the Internet. The difference between our feedback cyclic control policy and the DRR is that the upper bound for the service time is computed according to the arrival rate in our scheme while it is reserved in advance in [174]. Our method outperforms the DRR in the sense that no admission control nor resource reservations is required by our method.   □

Remark 7.4.2. It can known from (7.11) that the feedback cyclic control policy focuses on the case that the arrival rates are known in advance.   □

7.5 Comparison to Cyclic Fixed Interval Scheduling Method

We shall compare our algorithm with the *cyclic fixed interval algorithm* given in Example 1.1.1, which is widely used in practice. We consider the following three specifications:

- The idle time of the server excluding the set-up time.

- The convergence speed.

- The maximum buffer level after the system is stable.

For the sake of simplicity, consider the switched server system given in Example 7.3.1. To ensure that the system is stable, the fixed intervals of clients 1, 2 and 3 should be greater than 0.21, 0.42 and 1.26, respectively.

1. We shall first consider the idle time of the server excluding the set–up time. The experimental results for the *cyclic fixed interval algorithm* are given in Tables 7.4-7.6 where the fixed interval of each client is given in (   ).

   It can be known from the above tables that the idle time in the *cyclic fixed interval algorithm* is sometimes greater than the total service time while the idle time is always zero in our algorithm.

Table 7.4. The service time and the idle time of the server excluding the set–up time for the *cyclic fixed interval algorithm*

initial buffer level	service time (0.215,0.43,1.29)	idle time
(0, 0.5, 0.4)	4443.865	12761.905
(10,2.5,1.4)	7021.935	12324.065
(5,6.5,10.4)	10817.73	8449.27944

Table 7.5. The service time and the idle time of the server excluding the set–up time for the *cyclic fixed interval algorithm*

initial buffer level	service time (0.22,0.44,1.32)	idle time
(0, 0.5, 0.4)	954.4675	2567.973611
(10,2.5,1.4)	11208.8475	26158.9225
(5,6.5,10.4)	7718.8475	10943.366944

Table 7.6. The service time and the idle time of the server excluding the set–up time for the *cyclic fixed interval algorithm*

initial buffer level	service time (0.24,0.48,1.44)	idle time
(0, 0.5, 0.4)	4754.0375	14479.5325
(10,2.5,1.4)	5618.6175	14386.721389
(5,6.5,10.4)	5943.5375	133667.276944

2. We shall also compare the convergence speed between our algorithm and the fixed interval algorithm where the service time is defined by (7.4). We consider three different sets of initial values and the experimental results are given in Table 7.7-7.9 for the following three cases, respectively.

   a) $a = 0.9$, $b = 0.5$ and $c = 1.4$.

   b) $a = 10.0$, $b = 5.5$ and $c = 2.4$.

   c) $a = 5.0$, $b = 3.5$ and $c = 5.4$.

   Obviously, when Γ is small, the convergence speed of our algorithm is always faster than that of the *cyclic fixed interval algorithm*.

3. The maximum buffer levels after the system is stable are given in Table 7.10.

   Obviously, the maximum buffer levels of our algorithm are always less than those of the fixed interval algorithm. Thus, the total awaiting time of our algorithm is less than that of the fixed interval algorithm after the system is stable. Therefore, the proposed *feedback cyclic control policy* is overall better than the *cyclic fixed interval algorithm*.

Table 7.7. The convergence speed

Γ	our algorithm	cyclic fixed interval algorithm
0.1	324	702
0.5	87	134
0.8	66	80
1.0	59	63
2.0	46	27
3.0	42	15

Table 7.8. The convergence speed

Γ	our algorithm	cyclic fixed interval algorithm
0.1	2744	9610
0.5	571	1954
0.8	369	1218
1.0	301	973
2.0	167	482
3.0	122	319

Table 7.9. The convergence speed

Γ	our algorithm	cyclic fixed interval algorithm
0.1	1242	1802
0.5	271	354
0.8	181	218
1.0	151	173
2.0	92	82
3.0	71	52

Table 7.10. The maximum buffer level

Γ	our algorithm	cyclic fixed interval algorithm
0.1	[0.183,0.316,0.474]	[0.191,0.33.0.492]
0.5	[0.183,0.316,0.474]	[0.223,0.386,0.564]
0.8	[0.183,0.316,0.474]	[0.247,0.428,0.618]
1.0	[0.183,0.316,0.474]	[0.263,0.456,0.654]
2.0	[0.183,0.316,0.474]	[0.343,0.596,0.834]
3.0	[0.183,0.316,0.474]	[0.423,0.736,1.014]

7.6 Other Application Examples

Before we conclude this chapter, we shall briefly describe two other practical applications of the scheduling methods provided in this chapter. The details are left to the interested readers to explore.

The scheduling algorithm described in the previous section can be used to design an intelligent Left–Run and Right–Turn crossroads and Right–Run and Left–Turn crossroads as described in the following examples.

Example 7.6.1. **An Intelligent Crossroad Scheduling System**

Consider a crossroad system illustrated in Fig. 1.5, where A1, B1, C1 and D1 stand for Red–Amber–Green signals and A2, B2, C2 and D2 stand for Turn-Right (or Turn-Left) signals.

A crossroad scheduling system is said to be intelligent if it can measure the queuing length along each direction and determine the corresponding control law. Eight different groups of signals are

1 A1–Green, B1–Red, C1–Red, D1–Red,
   B2– Turn-Right, A2, C2, D2–No-Right-Turn;
2 A1–Amber, B1–Red, C1–Red, D1–Amber,
   A2, B2, C2, D2– No-Right-Turn;
3 A1–Red, B1–Red, C1–Red, D1–Green,
   A2–Turn-Right, B2, C2,D2–No-Right-Turn;
4 A1–Red, B1–Red, C1–Amber, D1–Amber,
   A2, B2, C2, D2–No-Right-Turn;
5 A1–Red, B1–Red, C1–Green, D1–Red,
   D2– Turn-Left, A2, B2,C2–No-Right-Turn;
6 A1–Red, B1–Amber, C1–Amber, D1–Red,
   A2, B2, C2, D2– No-Left-Turn;
7 A1–Red, B1–Green, C1–Red, D1–Red,
   C2–Turn-Left, A2, B2, D2–No-Right-Turn;
8 A1–Red, B1–Amber, C1–Red, D1–Red,
   A2, B2, C2, D2– No-Left-Turn.

The switchings of these processes are governed by the proposed improved scheduling method, that is

$$\psi(i) = i + 1 \; ; \; i = 1, 2, \cdots, 7, \tag{7.12}$$
$$\psi(8) = 1. \tag{7.13}$$

where $\psi()$ represents the switchings of the signals. □

With the development of sophisticated image processing techniques, the author believes that this type of intelligent crossroad systems will appear in the near future.

Similar idea can also be applied to design an intelligent bus-dispatch system as described in the following example.

Example 7.6.2. **An Intelligent Bus Dispatch System**

Consider a bus dispatch system for a bus line in a bus station.

A bus dispatch system is said to be intelligent if it can measure the queuing length and determine the corresponding scheduling law. Borrowing from the idea of the crossroad scheduling system, an intelligent bus dispatch system is proposed as follows:

1. If the queuing length is greater than a threshold, then a bus is set off immediately.

2. Otherwise, a bus is set off after a period.   □

8. Relative Differentiated Quality of Service of the Internet

In this chapter, the *feedback cyclic control policy* is first extended to the case that the arrival rates are not known in advance. The policy is then used to design a scheduling algorithm, called the *dual feedback cyclic control policy*, to provide a relative differentiated QoS in the current Internet. Meanwhile, a source adaptation scheme and an adaptive media playout scheme are proposed using the switched control for the Internet with the relative differentiated QoS. The overall structure is illustrated in Fig. 8.1. With this structure, a very low cost solution can be provided to transmit video over the Internet. This may be the first choice of many customers including students.

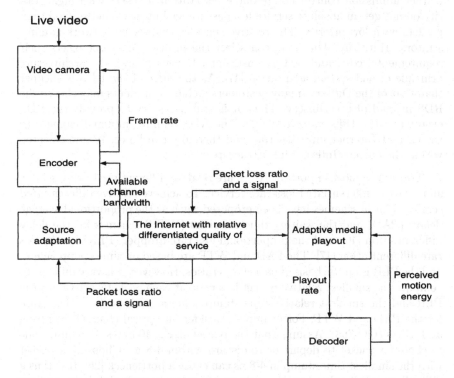

Fig. 8.1. The overall structure of our scheme

8.1 Quality of Service

The current Internet treats all packets with the same priority and without committing to any quality of service (QoS). However, many applications do require QoS, for example, video over the Internet [108, 110]. Therefore, the current Internet is not adequate for these applications. To solve this problem, the Internet Engineering Task Force (IETF) has defined two types of architectures: the integrated services (Intserv) [216] and the differentiated services (Diffserv) [39, 25]. The Intserv is the Internet incarnation of the traditional "circuit–based" QoS architecture and can provide a solid foundation for providing service with different priorities in the Internet. However, it mandates significant changes to the current Internet infrastructure. While the Diffserv keeps the core of the network as simple as possible and pushes most of the complexity to the network edges. Thus, compared to the Intserv, the Diffserv is highly scalable and relatively simple. We can foresee that it will dominate the backbone routers of the next generation Internet.

However, there is still a big gap between the current best effort service and the Diffserv. The most possible choice to upgrade the best-effort service is the relative differentiated service (RDS) [46]. In the RDS scheme, there is neither admission control nor resource reservation. Packets with high priority do not get an absolute service level assurance but just a better QoS than packets with low priority. The relative space can be set by network administrators arbitrarily. The users can select the priority that best meets their requirements, cost, and policy constraints. It thus follows the architecture principle of end-system adaptation [47]. As the price of the RDS is cheaper than that of the Diffserv, many customers including students may choose the RDS instead of the Diffserv. Thus, it is still necessary to provide the RDS even when the Diffserv is available. The RDS requires minimal changes to the current Internet infrastructure and therefore can be easily implemented within the current Internet infrastructure.

Two of the most important RDSs are relative differentiated delay service and relative differentiated loss-rate service. To achieve relative differentiated service, a lot of schemes have been proposed, such as the proportional average delays (PAD) and the waiting time priorities (WTP) schemes for the delay differentiation [46]; and the proportional loss rate dropper (PLR) for the loss rate differentiation [47]. The PAD and WTP are based on time stamp values, and the PLR is on the basis of packet loss ratios. However, it is very difficult to compute the smallest relative packet loss ratio at each packet's arrival for the PLR and the smallest relative time stamp value at each packet's departure for the PAD and WTP. For example, consider an optical channel operating at 40Gb/s (OC-768). Assume that the packet size is 40 bytes. It requires one to choose a packet to depart or to discard within 4-8 ns. Choosing a packet with the smallest time stamp in 4-8 ns can cause a bottleneck [26]. It is thus a difficult task to implement these schemes in a router with high link capacity.

8.2 Dual Feedback Cyclic Control Policy

The *feedback cyclic control policy* was proposed in the previous chapter to study the scheduling problem of a class of client/server systems modelled by equations (7.1) and (7.2) and the arrival rates are known in advance. However, each router in the Internet is usually modelled by Equation (1.7) and the arrival rates $f_i(k)(i = 1, 2, \cdots, n)$ are usually unknown.

The *feedback cyclic control policy* is extended to the case that the arrival rates are unknown, and is then applied to provide RDSs. The algorithm is called dual feedback cyclic control policy (DFCCP). In our proposed DFCCP scheme, the desired service rate of each priority traffic flow is first computed using the fluid flow traffic model (1.7) according to the desired buffer delay of each priority independently. The total bandwidth CT is allocated to each priority proportionally according to their desired service rates. Let $u_i(k)$ denote the bandwidth allocated to priority i. At the kth time interval, there are exact $u_i(k)$ of bits that will be sent out for priority i. To minimize the buffer delay of each packet and to provide the relative differentiated services, a time varying sending quantum and a time varying discarding quantum are computed for each priority by taking the maximum transfer unit (MTU) into consideration. The sending quantum and discarding quantum, similar to $S_j(k)$ in Equation (7.11), are the upper bounds of the service time within the kth interval. Two *feedback cyclic control policies* (FCCPs) are then provided to achieve the relative differentiated services: one for the relative differentiated delay service and the other for the relative loss-rate service.

In the FCCP for sending packets, the buffer occupancy is a monitored parameter. When the buffer occupancy of a priority increases, it means that the bandwidth allocated to the priority is less than its arrival rate. To maintain the desired buffer delay, the service rate should be increased at the next time slot. Conversely, the service rate will be reduced at the next time slot if the buffer occupancy decreases.

Obviously, the FCCP can work well when the total load is less than 100%. However, in case of overload, the router cannot send out all the arriving packets within each time slot. The remaining packets will be kept in the buffer. The buffer occupancy will keep on increasing and so does the delay. This is not consistent with the principle of the FCCP for sending packets and so it must cooperate with a dropping scheme. In the FCCP for dropping packets, a dynamic threshold is computed according to the maximum tolerable buffer delay to detect congestion. The dropping quantum of each priority is determined by the desired service rate and the relative packet dropping factors, which would be used to discard packets in case of congestion to achieve the relative differentiated packet dropping.

The DFCCP scheme is very simple in the sense that it only computes the sending quantum and discarding quantum at every time slot T and does not need the time stamp value. Hence, it reduces the requirement of high computation capability and can be easily implemented on a router with high link capacity.

8.2.1 The Service Rate

The structure of router is shown in Fig. 1.2. It has an output link with transmission capacity Cbits/s and a fixed buffer space B_s bits. The router is non-preemptive, which means that a packet transmission is always carried out to completion. However, the router is not work-conserving such that the jitter is somewhat controllable. The work-conserving property implies that the router does not stay idle when there are waiting packets and the operation of the router can be as described below.

An arriving packet is queued based on its priority marking in one of these queues. An enqueuing module monitors the aggregate backlog of waiting packets in the forward engine, and computes a service rate for each priority according to the desired buffer delay at the beginning of each period with length T. The sending quantum and discarding quantum are then calculated as the upper bounds for two *feedback cyclic control policies*: one is used to determine which packet should be served from the head next, and the other one is applied to determine which packet should be discarded in case of congestion. It should be highlighted that these two FCCPs may be applied many times within each period with length T so as to achieve the necessary QoS differentiation among priorities.

In our scheme, the system will compute the service rate at every T units of time, i.e. it collects system information and adjusts the parameters at each of time instant kT $(k = 0, 1, \cdots)$.

Suppose that there are a total of n priorities of services, with priority 1 being the highest and n the lowest. In the DFCCP, the service rates $\hat{u}_i(kT)(i = 1, 2, \ldots, n)$ are calculated by [107]

$$\hat{u}_i(k) = \max\{0, u_i(k-1) - X_i(k-1) + 2X_i(k) - \frac{v_i(k-1)\gamma_i}{T}\}, \tag{8.1}$$

$$\hat{u}_i(0) = \frac{CT}{2n}, \tag{8.2}$$

where γ_i is the desired buffer delay of priority i, and satisfying

$$\gamma_n \geq \gamma_{n-1} \geq \cdots \geq \gamma_1. \tag{8.3}$$

and $\gamma_i(i = 1, 2, \cdots, n)$ can be arbitrarily set by network administrators, T is the length of the time slot for the computation of service rates, and its value is recommended to be $2^{-\ell}$ where ℓ is an integer, and $v_i(k-1)$ is the "average" value of $\hat{u}_i$ that is computed by a sliding window method given by

$$v_i(k-1) = \frac{\hat{u}_i(k-1)}{2^\ell} + \frac{(2^\ell - 1)v_i(k-2)}{2^\ell}, \tag{8.4}$$

$$v_i(0) = \frac{CT}{2n}, \tag{8.5}$$

From our experimental experience $\ell = 5$ is a suitable value.

For simplicity, assume that the maximum transfer unit for all priority packets MTU is known in advance and $CT > n * MTU$. The service rate is adjusted by

$$\tilde{u}_i(k) = \begin{cases} \hat{u}_i(k) & \text{if } \hat{u}_i(k) \geq MTU \\ MTU & \text{if } MTU > \hat{u}_i(k) \geq \theta MTU \\ 0 & \text{otherwise} \end{cases}, \tag{8.6}$$

$$\theta = 1 - \min\{1/2, \frac{\sum_{j=1}^{n} \hat{u}_j(k)}{CT}\}. \tag{8.7}$$

Different service rates are computed for packets with different priorities based on their desired buffer delay. However, the sum of $\tilde{u}_i$ may be greater than the total transmission capacity CT. Therefore, a relative fair share rate should be allocated to packets with each priority according to the following two cases:

Case 1. When $\sum_{i=1}^{n} \tilde{u}_i(k) \leq CT$, the router is under loaded. The final sending rate is set as $\tilde{u}_i(k)$.

Case 2. Otherwise, the overall load is very heavy. The final service rate is set as

$$u_i(k) = \begin{cases} MTU, & \text{if } \tilde{u}_i(k) = MTU \\ \dfrac{(CT - \displaystyle\sum_{j,\tilde{u}_j(k)>0} MTU)\tilde{u}_i(k)}{\sum_j \tilde{u}_j(k)} + MTU, & \text{if } \tilde{u}_i(k) > MTU \\ 0, & \text{otherwise} \end{cases}. \tag{8.8}$$

8.2.2 Feedback Cyclic Control Policy for Sending Packet

An upper bound of the sending quantum allocated to queue i with positive $u_i(k)$, $Q_i(k)$, in the kth period with length T, is computed as

$$Q_i(k) = \begin{cases} \frac{u_i(k)MTU}{u_{i_0}(k)} & \text{if } i \neq i_0 \\ MTU & \text{otherwise} \end{cases}, \tag{8.9}$$

where $i_0 = \arg\min_{j,u_j(k)>0}\{u_j(k)\}$. It can be easily shown that $u_i(k) \geq Q_i(k) \geq MTU$.

Let us first define the following symbols used in the sending algorithm:

queue i- the ith queue, which stores packets with priority i;
enque(), deque() - standard queue operators;
DC_i - contains the bytes that queue i can used in the recent round;
$LeftDC_i$ - contains the bytes that queue i can used in the recent period;
$sendinglist$ - the list of queues for sending packets.

The FCCP algorithm for sending packets is described below.

Algorithm 8.1 *Feedback cyclic control policy for Sending Packets*

Initialization:
For each queue:{
$DC_i = 0;$
$LeftDC_i = 0;$
$Q_i(k) = 0;}$

Enqueuing module:
On arrival of packet p
$i = ExtractFlow(p);$
enque(i,p);

Allocate throughput:
At time kT
Set the sendinglist as empty;
For (each queue){
If (Empty(queue i)){
$DC_i = 0;$
$LeftDC_i = 0;}}$
Compute $u_i(k)$ by (8.1)-(8.8);
If($u_i(k) > 0$){
$LeftDC_i(k) = u_i(k)$;
Compute $Q_i(k)$ by (8.9);
Add i to the sendinglist;}

Dequeuing module:
While (TimeSlot(k)) {
While (sendinglist is not empty){
Remove head of sendinglist, say flow i{
Compute the sending quantum by
$\hat{Q}_i(k) = Q_i(k).$
If(LeftDC$_i \geq \hat{Q}_i(k)$){
$DC_i = DC_i + \hat{Q}_i(k);$
$LeftDC_i = LeftDC_i - \hat{Q}_i(k).}$
Else{
$DC_i = DC_i + LeftDc_i;$
$LeftDC_i = 0.}$
Flag=1;

While((Flag=1)or ((Flag=0) and ($DC_i > 0$)) and (queue i is not empty){
PacketSize = Size(Head(queue i));
If((Flag=1) or ((PacketSize < DC_i) and (Flag=0))) {
send deque(queue i);
$DC_i = DC_i - PacketSize;}$

Else
break;
Flag=0;}
If (flow i still active)
Add i to the end of sendinglist;
Else
Delete queue i from sendinglist;
}//end of Remove head of sendinglist, say flow i
}//end of While (sendinglist is not empty)
}//end of While (TimeSlot(k))

The FCCP is based on the observation that the desired service rate somewhat matches the arrival rate when the load is less than 100%. The desired service rate is thus used to determine the desired buffer occupancy. In this way, the FCCP can control the delay of each priority.

However, when the load is more than 100%, the service rate cannot match the arrival rate and so the FCCP dose not work. Hence, a dropping scheme is needed to discard packets when necessary.

8.2.3 Feedback Cyclic Control Policy for Dropping Packet

1. Congestion–Detector The congestion-detector must be adjusted by the desired service rate. Also, it should be able to tolerate that the buffer occupancy can be greater than the desired buffer occupancy to a certain extent. So we introduce two parameters $\beta(\beta > 1)$ and $\hat{\beta}$, and propose a dynamic threshold as

$$Td(kT) = \max\{\sum_{i=1}^{n} \frac{v_i(kT)}{T}\beta\gamma_i, \sum_{i=1}^{n} \hat{\beta}C\gamma_i\}. \tag{8.10}$$

The function of β and $\hat{\beta}$ is to obtain a good tradeoff between the packet loss ratio and the buffer delay.

The congestion is detected if the total buffer occupancy satisfies

$$\sum_{i=1}^{n} X_i > Td(kT). \tag{8.11}$$

In equation (8.10), $\beta\gamma_i$ can be regarded as the maximum tolerable delay for packets.

2. Packet–Dropper

The FCCP for dropping scheme first computes a discarding quantum for each priority. In case of congestion, packets are dropped from the tail of each queue with the push-out technique. In other words, *packets in the tail of each queue instead of the coming packets are dropped in our scheme*, until it uses up all the quantum. Then a new discarding quantum is added to each priority. The discarding quantum is updated (just updated, not added) at every time slot, together with the desired service rate.

The discarding quantum allocated to $queue_i$ with $u_i(kT) > 0$, i.e. $\tilde{Q}_i(kT)$, in the kth period, are then computed as

$$\tilde{Q}_i(kT) = \frac{MTU}{u_{j_0}(kT)\theta_{j_0}} u_i(kT)\theta_i, \tag{8.12}$$

where $j_0 = arg\min_i\{u_i(kT)\theta_i\}$, $\theta_1 \leq \theta_2 \leq \cdots \leq \theta_n$ with θ_i being used to control the different drop rate of priorities and be arbitrarily set by network administrators, and $u_i(kT)$ are the desired service rate computed in the FCCP for sending packets.

We define other symbols used in our algorithm as

$queue_i$ – the ith queue, which stores packets with priority i.
discard(), pushout() – standard queue operators.
$DCdrop(i)$ – contains the bytes that can be used for now.
$Discardinglist$ – the list of queues which still can drop packets.

The relative differentiated dropping scheme is then presented as

Algorithm 8.2 *Feedback cyclic control policy for Dropping Packets*

> **Initialization:**
> **For** *(each queue):* {
> $DCdrop(i) = 0;$
> }
>
> **At time** kT
> **Compute** $DCdrop(i):$
> *Set the Discardinglist as empty.*
> **For** *(i = n;i > 0;i − −)* {
> **If***(u_i(kT) > 0)*
> *Compute* $\tilde{Q}_i(kT)$ *by (8.12);*
> *Add i to the end of the Discardinglist;*
> }
> }
>
> **When congestion is detected, drop packets***:*
> **While** *(congestion)* {
> **If** *(all DCdrop(i) < MTU)* {
> *Set Discardinglist empty;*
> **For***(i = n; i > 0; i − −)* {

$DCdrop(i) = DCdrop + \tilde{Q}_i(kT);$
Add i to the end of Discardinglist;
}
}
While *(Discardinglist is not empty) and (congestion) {*
Remove head of Discardinglist, say flow i;
While *(DCdrop(i) > 0) and (queue$_i$ is not empty) and (congestion)*
{
PacketSize=size(tail(queue$_i$));
If *(PacketSize < DCdrop(i)) {*
discard(pushout(queue$_i$));
$DCdrop(i) = DCdrop(i) - PacketSize;$
}
Else
break;
}
If *(flow i is still active)*
move flow i to the end of Discardinglist;
Else
delete flow i from Discardinglist;
}
}
}

Three examples with 3 priority traffic are given: Example 8.2.1 is to check the performance of DFCCP with fixed traffic (stationary load distribution); Example 8.2.2 is to show the efficiency of DFCCP with dynamic traffic; and Example 8.2.3 is to show the effect on the choice of the parameters in DFCCP. In all examples, the arrival of packets obeys Poisson distribution and the size is uniformly distributed.

Example 8.2.1. **DFCCP with Fixed Traffic Load**

Four scenes are considered and the overall traffic loads are set at 60%, 70%, 80% and 90%, respectively. In each scene, traffic load is assigned to each priority with ratios given in Table 8.1. Table 8.2 contains parameters used in the experiment. Table 8.3 summarizes the average delay and loss rate of each priority. It indicates that the DFCCP can control the traffic flow more accurately in case of heavy load. Table 8.4 shows the result of FCCP sending algorithm working with a PLR(∞) dropping scheme. The same congestion detector is used and the parameters are set as those in Table 8.2. Fig. 8.2 illustrates the comparison of packet loss ratios, which shows that our proposed DFCCP has provided a different delay service and a different loss service. □

Table 8.1. Traffic load assignment

case	priority 1	priority 2	priority 3
1	1	2	4
2	1	1	1
3	1	$\frac{1}{2}$	$\frac{1}{2}$

Table 8.2. Parameters sets

sets	γ_1	γ_2	γ_3	β	θ_1	θ_2	θ_3
1	0.001	0.002	0.003	2	1	2	6

Table 8.3. DFCCP with parameters sets 1

load	case	delay($\times 10^{-6}$ sec)			Loss rate (%)		
		P. 1	P. 2	P. 3	P. 1	P. 2	P. 3
	1	547	927	1583	0.0370	0.0765	0.2222
60%	2	391	896	1562	0.0979	0.1820	0.5262
	3	317	889	1677	0.2711	0.4919	1.3722
	1	669	1108	1908	0.0283	0.0541	0.1692
70%	2	495	1079	1846	0.0804	0.1547	0.4327
	3	399	1073	1918	0.2421	0.4379	1.2144
	1	834	1387	2355	0.0239	0.0459	0.1413
80%	2	669	1376	2263	0.0741	0.1428	0.4209
	3	559	1363	2259	0.2250	0.4104	1.1532
	1	1102	1849	2998	0.0292	0.0622	0.1894
90%	2	1005	1868	2861	0.0929	0.1849	0.5580
	3	890	1856	2761	0.3334	0.6323	1.8168

Table 8.4. PLR with parameters sets 1

load	case	delay($\times 10^{-6}$ sec)			Loss ratio (%)		
		P. 1	P. 2	P. 3	P. 1	P. 2	P. 3
	1	544	922	1580	0.0387	0.0753	0.2379
60%	2	389	892	1556	0.0860	0.1685	0.5182
	3	317	887	1654	0.2479	0.4900	1.4940
	1	668	1102	1911	0.0259	0.0491	0.1502
70%	2	491	1082	1843	0.0675	0.1347	0.4160
	3	402	1076	1951	0.2211	0.4410	1.3245
	1	821	1375	2351	0.0246	0.0455	0.1415
80%	2	667	1375	2267	0.0675	0.1353	0.4083
	3	559	1364	2253	0.2252	0.4433	1.3371
	1	1093	1842	2984	0.0300	0.0590	0.1801
90%	2	996	1863	2859	0.0972	0.1967	0.6013
	3	883	1849	2766	0.3009	0.5981	1.7967

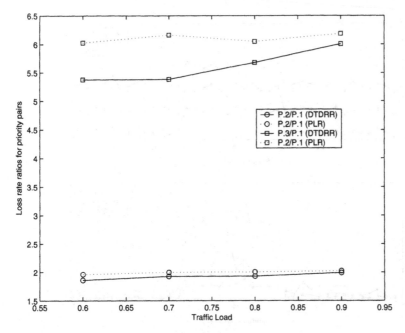

Fig. 8.2. The ratio of loss rate for priority pairs

Example 8.2.2. **DFCCP with Dynamic Traffic Load**

Traffic load is first set as 95%, then switched to 105% and finally switched back to 95%. Fig. 8.3 illustrates the result of DFCCP; Fig. 8.4 demonstrates the result with PLR(∞). Notice that, when the system is overloaded, the delay of each traffic is increased. This is understandable since our DFCCP tries to obtain a good tradeoff between the packet loss ratio and the buffer delay. In practice, the maximal delay can be adjusted by setting a suitable β.   □

Example 8.2.3. **Effect of Parameter of the DFCCP**

In this example, the effect on the choice of β is checked. From the principle of DFCCP, we know that β will affect the total number of loss. In other words, the total loss rate is reduced with an increase in β. Fig. 8.5 illustrates the average loss rate with different β under different traffic load condition. Fig. 8.6 shows the average loss rate versus traffic load with a certain β.   □

It is clear from these examples that the proposed scheme is consistent with the FCCP sending scheme to achieve the differentiated dropping rate for packets with different priorities. The FCCP dropping scheme can achieve the same performance as the PLR(∞) scheme with the same congestion detector. However, our proposed FCCP dropping scheme is much simpler than the PLR(∞) scheme.

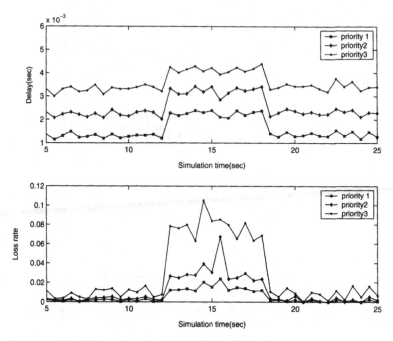

Fig. 8.3. DFCCP with switched traffic load

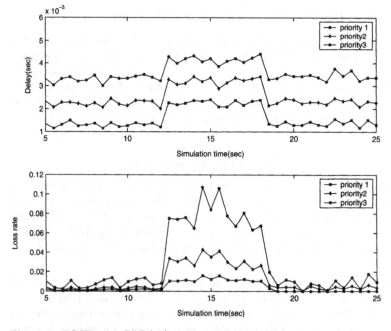

Fig. 8.4. FCCP with PLR(∞) with switched traffic load

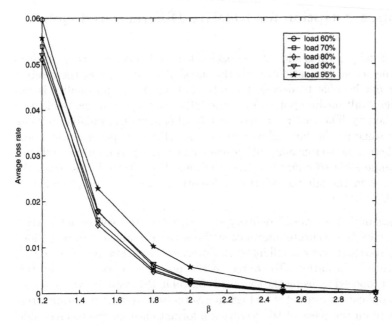

Fig. 8.5. Average loss rate vs β with same traffic load

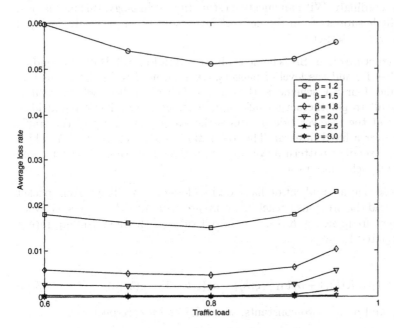

Fig. 8.6. Average loss rate vs traffic load with same β

8.3 Source Adaption for Relative Differentiated QoS

There is usually a threshold for the packet loss ratio over a cluster of video data. If the actual loss ratio exceeds the threshold, a decoder at the receiver side may not be able to decode the bitstream or the visual quality is unacceptable. Multimedia applications especially video applications tend to be resource-hungry. They are ready to consume all bandwidth available to them. However, same as the best effort service, the RDS does not guarantee any QoS. If there is no source adaptation scheme in place, it is then very difficult or even impossible to guarantee that the packet loss ratio is less than the threshold, even though the relative differentiated QoS can be provided by our DFCCP [189, 14].

To maximize the overall delivery over network, the source adaptation prefers to have the accurate information about the available bandwidth along the path, averaged over a small interval. However, the current network doesn't provide such information. To overcome the problem, a source adaptation scheme is proposed by using the switched control in this section. At the sender side, an application evaluates and estimates the service that it obtains from the network on the basis of the feedback information from the receiver side, and adjusts its sending pattern including the sending rate and the level of packets accordingly. With an effective estimation technology, this mechanism requires little support from the network layer, so it is can be easily deployed on the current Internet.

The structure of our SA scheme is illustrated in Fig. 8.7. Both the sending rate, $K(k+1)$, and the level of packets, $l(k+1)$, need to be determined by our scheme. Our SA scheme is thus a switched controller [30]. Since it is very difficult to get the actual information about the available bandwidth, a counter at the receiver side measures the average loss ratio of the traffic packages over a fixed interval. The loss ratio is sent back to the SA, which adjusts the sending pattern according to the trend of packet loss ratio and the current packet loss ratio.

Suppose that an application has a desired loss ratio γ_1. The current packet loss ratio and the previous packet loss ratio are $\zeta(k)$ and $\zeta(k-1)$, respectively. The current frequency is $K(k)$. The initial value of the next sending rate is then computed by [30]

$$\hat{K}(k+1) = K(k)(1 + \alpha_p(\gamma_1 - \zeta(k)) + \alpha_d(\zeta(k-1) - \zeta(k))), \qquad (8.13)$$

where α_p and α_d are two constants, set as 0.8 and 0.2, respectively.

If the traffic loss ratio is higher than the desired one, then the source reduces the size of each frame, hence it reduces the load of network. Otherwise, the source increases the size of each frame to achieve a higher utilization of the available bandwidth.

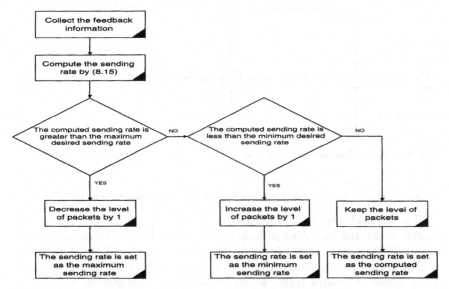

Fig. 8.7. The overall structure of our source adaptation scheme

To minimize the overall cost, it is desirable to use the lowest level to satisfy the requirements. Our scheme is then described by

$$K(k+1) = \begin{cases} K_{min} & \text{if } \hat{K}(k+1) < K_{min} \\ K_{max} & \text{if } \hat{K}(k+1) > K_{max} \text{ ,} \\ \hat{K}(k+1) & \text{otherwise} \end{cases} \qquad (8.14)$$

and

$$l(k+1) = \begin{cases} l(k) + 1 & \text{if } \hat{K}(k+1) < K_{min} \\ l(k) - 1 & \text{if } \hat{K}(k+1) > K_{max} \text{ ,} \\ l(k) & \text{otherwise} \end{cases} \qquad (8.15)$$

where K_{min} and K_{max} are the lower bound and the upper bound of $K(k)$.

Both K and l are used by the application as the default frame encoded frequency and the default level until they are replaced by new values.

Clearly, our scheme tries to use the lowest (also the cheapest) level to meet the desired relative differentiated QoS. Moreover, our scheme provides a very smooth rate control in the sense that the ratio of two successive values is near to 1, which can be well matched to real time applications over the Internet. It does not depend on the accurate measurements of the connect trip time.

Two examples are given below to verify our proposed SA scheme. A video application is used as an example to test our scheme.

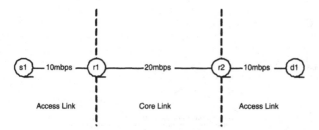

Fig. 8.8. Single link network topology

Example 8.3.1. **Source Adaption for the Best Effort Service**

The network topology is illustrated in Fig. 8.8, and it consists of two routers connected by a 20mbps link. One video session is established to send video packets form s1 to d1. To create different network condition, data source is attached on r1 to create traffic from r1 to r2. The data source is a simple UDP source, it creates traffic which obeys Possion distribution and has no adaptation mechanism. When the core link provides a relative differentiated service, there will be multiple data sources and each source feeds a base traffic to a service level. There is no traffic load on the path from d1 to s1, so the feedback information of video session will not be lost. The video is encoded at 30f/s, while K_{min} and K_{max} are set as 540 and 3125, respectively. Thus, the minimal and maximal sending rates of video session are $128kbps$ and $0.75mbps$, respectively. This is a typical scope of video traffic over the current Internet. And the packet size is 150 bytes.

Network provides a first in first out (FIFO) service, instead of the differentiated quality of services. The objective is to evaluate the effect of adjusting sending rate within a level. Upon the core link, a heavy base traffic is loaded and leads to a limited available bandwidth. The desired loss rate of video session is set at 1%. Figs. 8.10-8.11 show the sending rate, the dropping rate and the utilization of core link versus available bandwidth of the core link, respectively. With an increase in the base traffic load, the available bandwidth decreases. Our proposed scheme can adjust the sending rate correspondingly and make full use of the bandwidth while keeping the loss rate of video session in desired level when the base traffic load is less than 90%. However, when the base traffic load is greater than 95%, our source adaptation scheme does not work any more. This is due to the lower bound of the sending rate K_{min} and insufficient resource. □

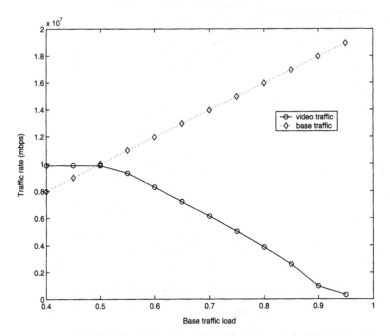

Fig. 8.9. The sending rate with different available bandwidth

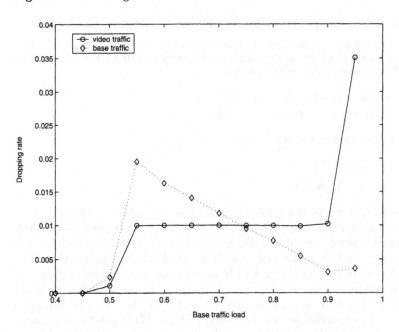

Fig. 8.10. The dropping rate with different available bandwidth

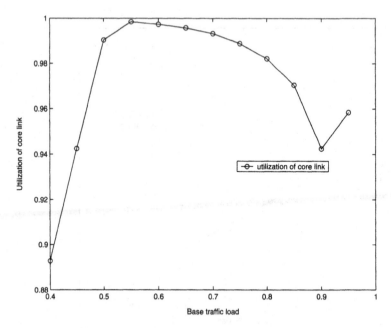

Fig. 8.11. The utilization of core link with different available bandwidth

Example 8.3.2. **Source Adaption for Relative Differentiated QoS**

The setting of this example is the same as that of Example 8.3.1. Network provides the relative differentiated service by the DFCCP with 6 service levels. The desired delay of each level is

$$\gamma_1 = 0.001s \; ; \; \gamma_2 = 0.002s \; ; \; \gamma_3 = 0.003s \; ;$$
$$\gamma_4 = 0.004s \; ; \; \gamma_5 = 0.005s \; ; \; \gamma_6 = 0.006s.$$

The drop rate parameters $\theta_i (i = 1, 2, \cdots, 6)$ are

$$\theta_1 = 1 \; ; \; \theta_2 = 2 \; ; \; \theta_3 = 3 \; ;$$
$$\theta_4 = 4 \; ; \; \theta_5 = 5 \; ; \; \theta_6 = 6.$$

In the core link, there is a base traffic load of up to 99%, with an average available bandwidth $0.2mbps$; Fig. 8.12 shows the base traffic of each level. The desired loss rate of video traffic is 1%. With the DFCCP algorithm, the average delay and dropping rate of each service level are illustrated in Figs. 8.13 and 8.14, respectively. Fig. 8.15 illustrates the rate of video traffics, and Fig. 8.16 demonstrates the service level taken by the video traffic. It should be mentioned that in such a load case, it is only with the RDS that the video traffic can get a satisfied service. However, with the FIFO scheme, it is shown in Fig. 8.10 that the loss rate is around 1.3% when the base traffic load is less than 90% and reaches 4% when the base traffic load is 99%. This implies that the video application will work poorly with the FIFO service. This implies that our DFCCP is very useful for real time applications.   □

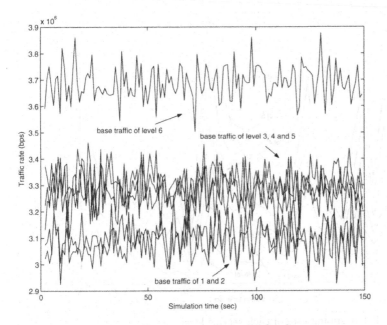

Fig. 8.12. The base traffic rate on each service level

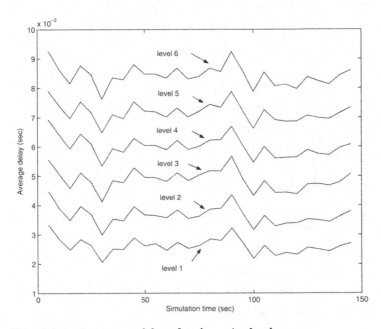

Fig. 8.13. The average delay of each service level

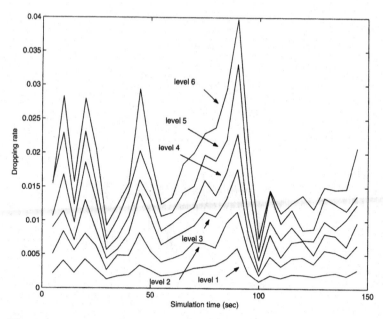

Fig. 8.14. The dropping rate of each service level

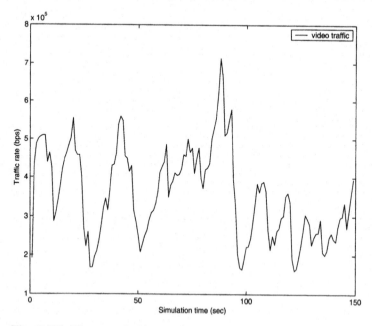

Fig. 8.15. The rate of video traffic

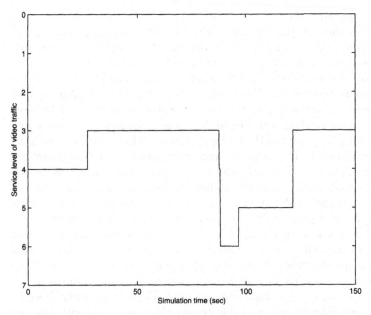

Fig. 8.16. The service level of video traffic

It is thus shown that the proposed source scheme can adjust application according to network situation when the relative differentiated quality of service is available in the Internet.

8.4 Adaptive Media Playout for Relative Differentiated QoS

Besides the SA scheme at the sender side, it is also desirable to provide an adaptation scheme at the receiver side. This section focuses on the adjustment at the receiver side. Particularly, we study an adaptive media playout (AMP) scheme to improve the robustness of real time video applications over the Internet with RDS. The AMP is adopted to reduce both pre-roll delay and viewing latency of video applications when it is delivered from a remote server. The pre-roll delay is the time to fill a client buffer to a desired threshold so that the playout rate can start after a user's request, while the viewing latency is the time interval separating a live event and its viewing time at the client.

The AMP is an efficient receiver-driven and delay based method for the playout rate control. During the burst loss, no further packet is received and the number of frames in the buffer falls below a given threshold, the playout rate is reduced. After the burst loss is over, the buffer level will begin to increase, after the buffer occupancy reaches the threshold, the video is played at the normal playout rate. If the buffer occupancy continues to increase and the buffer occupancy is above the threshold, the playout rate is further increased. The AMP can thus be used to increase the robustness of the playout process with respect to adverse channel conditions and it was studied by many researchers. Yuang et al [214] proposed an AMP scheme which employs a maximum play-out rate U when the number of buffered frame $B(k)$ is greater than a threshold N_{adapt} and reduces the playout rate at $\mu(k) = UB(k)/N_{adapt}$ when the buffer occupancy is less than the threshold. Since the arrival rate is unknown, the maximum playout rate may be greater than the arrival rate. As a result, the playout rate sometimes may be too low and the visual quality is unacceptable. To overcome this, Kalman et al. [80] proposed another AMP scheme and the AMP playout time is defined as κ_1/F_r, $1/F_r$ or κ_2/F_r with a coded frame rate F_r, a slowdown factor $\kappa_1 > 1$ and a speed-up factor $\kappa_2 < 1$. The scheme works very well when the number of incoming packets varies slowly. However, the playout rate changes too frequently when the number of incoming packets varies a lot and quickly. The visual quality is also unacceptable in this case. Moreover, the slowdown factor and the speedup factor are fixed. This is not optimal because human eyes have different tolerance degree over video sequences with different motions.

In this section, we propose an AMP scheme to solve the problem associated with the existing AMP schemes [214, 80]. Our scheme is based on the human visual system (HVS). Human eyes are more sensitive to high motion video sequence, perceived motion judder is very serious if the playout rate is not properly adjusted. However, this may cause no annoyance with the same adjustment when low motion is presented. Thus, different upper and lower bounds should be defined for the playout rate according to the perceived motion energy of video sequences. Meanwhile, if the playout rate is not controlled well, the buffer may be underflowed for a period, and certain frame will be repeated within the period. Obviously, the visual quality will not be good. A simple control theory is adopted in our scheme to compute the playout rate by taking both the trend of buffer occupancy variation and the difference between the buffer occupancy and the threshold into consideration. The playout rate is further bounded by the above two bounds. Our scheme is thus a motion based AMP scheme.

8.4.1 The Structure of Our Proposed Adaptive Media Playout Scheme

The AMP enables the client to control the playout rate of video by simply scaling the durations of those frames that are shown. The short-term dips in the channel capacity can be absorbed with acceptable distortion on the viewing latency. The AMP is thus able to improve the robustness of the playout process with respect to adverse channel conditions.

In our proposed motion based AMP scheme, as illustrated in Figure 8.1, a playout rate is first computed by considering both the trend of buffer occupancy variation and the difference between the buffer occupancy and a predefined threshold. The PME of a video sequence is calculated and is used to determine the dynamic upper and lower bounds for the playout rate. Because of possible poor channel conditions, a special request will be sent to the SA to increase the priority of video packets when the buffer is in danger of underflow. When the buffer occupancy is above the threshold, a signal will be delivered to the sender to indicate that the source adaption can determine the priority of packets only according to the network condition [30].

8.4.2 Perceived Motion Energy

The perceived motion energy (PME) of a frame is introduced to indicate the degree of perceived motion judder if the playout rate is adjusted. The PME is defined based on the following observation:

Because of the limited resources, the HVS devotes energy to the measurement of local direction and the speed of motion.

In the H.264 bitstream [192], there are zero, one or two motion vectors in each block with size 4x4 of P-frame or B-frame for motion compensation, often referred to as the motion vector field (MVF). In the MVF, let (i, j) be the position of blocks in a raster scan order in the kth frame, (dx, dy) be the motion vector (MV) of the block, and the temporal distance from the reference frame to the current frame be $\rho_{i,j}(k)$. The normalized MV is

$$\bar{dx}_{i,j}(k) = \frac{dx}{\rho_{i,j}(k)} \; ; \; \bar{dy}_{i,j}(k) = \frac{dy}{\rho_{i,j}(k)}. \tag{8.16}$$

The energy $En_{i,j}(k)$ is defined as

$$En_{i,j}(k) = \bar{dx}_{i,j}^2(k) + \bar{dy}_{i,j}^2(k). \tag{8.17}$$

The second moment of motion energy instead of the mean value is used to give more weight to the high motion because human eyes are more sensitive to the direction with high motion.

The angle 2π is quantized into 8 ranges, and the number of motion vectors in the nth $(n = 1, 2, \cdots, 8)$ range is denoted by $AH(k, n)$. The total energy of MVs in the nth $(n = 1, 2, \cdots, 8)$ range is denoted by $AE(k, n)$ and is given by

$$AE(k, n) = \sum_{MB_{i,j} \in \text{range } n} En_{i,j}(k). \tag{8.18}$$

The average values of $AH(k,n)$ and $AE(k,n)$ over a window of size 4 are

$$\bar{AH}(k,n) = (AH(k,n) + 3*\bar{AH}(k-1,n))/4, \tag{8.19}$$
$$\bar{AE}(k,n) = (AE(k,n) + 3*\bar{AE}(k-1,n))/4. \tag{8.20}$$

We then define the dominant motion direction n_0 as

$$n_0 = \text{argmax}_l\{\varphi(\bar{AH}(k,l),\bar{AE}(k,l)), l \in \{1,2,\cdots,8\}\}, \tag{8.21}$$

where $\varphi(\bar{AH}(k,l),\bar{AE}(k,l))$ is a polynomial function of $\bar{AH}(k,l)$ and $\bar{AE}(k,l)$.

The perceived motion energy is defined as

$$PME(k) = \frac{\bar{AE}(t,n_0)}{\sum_{n=1}^{8} \bar{AH}(k,n)} = \frac{\bar{AE}(k,n_0)}{\bar{AH}(k,n_0)} \frac{\bar{AH}(k,n_0)}{\sum_{n=1}^{8} \bar{AH}(k,n)}.$$

Clearly, the PME is the product of the average value of motion energy and the percentage of dominant motion direction. The first item is used to indicate that the adjustment of frame with low motion is less perceptible than that with high motion. The second item is used to simulate the characteristics of human perception that human's eyes tend to track dominant motion in the scene. Let the just-noticeable-distortion of frame rate for the kth frame be denoted by $JND_{fr}(k)$ and it is satisfied that

$$JND_{fr}(k) = \frac{\zeta(\frac{\bar{AE}(k,n_0)}{AH(k,n_0)}, \frac{\bar{AH}(k,n_0)}{\sum_{n=1}^{8} AH(k,n)})}{\frac{AE(k,n_0)}{AH(k,n_0)}}, \tag{8.22}$$

where $0 < \frac{\partial \zeta(x,y)}{\partial x} < 1$ and $\frac{\partial \zeta(x,y)}{\partial y} < 0$.

Predefine two thresholds θ_1 and θ_2. The sequence is said to be with high motion if $PME(k) \geq \theta_1$, and be with medium motion if $\theta_1 > PME(k) \geq \theta_2$. Otherwise, it is said to be with low motion. θ_1 and θ_2 are selected as 16 and 4, respectively.

8.4.3 The Playout Rate

Let $B(k)$ denote the number of continuous frames in the AMP buffer. The dynamics of $B(k)$ is modelled as

$$B(k+1) = \max\{0, B(k) + I(k) - \mu(k)\}, \tag{8.23}$$

where $I(k)$ and $\mu(k)$ are the arrival rate and the playout rate, respectively.

Assume that the original video is coded at a rate of F_r. Packets are removed from the palyout queue at a rate of $\mu(k)$. N_{adapt} is a predefined threshold value. The initial vale of the next playout rate is then computed by [30]

$$\hat{\mu}(k+1) = \mu(k) + \frac{(B(k) - N_{adapt}) + 15 * (B(k) - B(k-1))}{16}, \tag{8.24}$$

$$\hat{\mu}(0) = F_r. \tag{8.25}$$

Obviously, the playout rate is adjusted very smoothly by our scheme.

Let $STT(k)$ denote the single trip time from the sender to the receiver. When the buffer is in danger of underflow, i.e. $B(k) \leq 2.5*STT(k)*(7F_r/8 - (B(k)-B(k-1)+\mu(k)))$, a special request is sent to the sender to increase the priority of video packets. When $B(k) > N_{adapt}$, a signal will be delivered to the sender to indicate that the SA can adjust the priority of packets according to the network conditions.

Since human eyes have different sensitivity to video with different motions, content adaptive upper and lower bounds, $L(k+1)$ and $U(k+1)$, are computed based on content for the final palyout rate. The final visual quality of the play out video sequence can then be guaranteed. It can be known from (8.22) that the variation of frame rate for frame with low PME should be larger than that with high PME. $L(k+1)$ and $U(k+1)$ are then accordingly given by

$$L(k+1) = \begin{cases} \frac{7F_r}{8} & \text{high motion} \\ \frac{3F_r}{4} & \text{medium motion} \\ \frac{5F_r}{8} & \text{low motion} \end{cases}, \tag{8.26}$$

$$U(k+1) = \begin{cases} \frac{9F_r}{8} & \text{high motion} \\ \frac{5F_r}{4} & \text{medium motion} \\ \frac{11F_r}{8} & \text{low motion} \end{cases}. \tag{8.27}$$

The initial values of $L(k)$ and $U(k)$ are $3f_r/4$ and $5F_r/4$, respectively.

The playout rate is bounded by

$$\tilde{\mu}(k+1) = \max\{L(k+1), \min\{U(k+1), \hat{\mu}(k+1)\}\}. \tag{8.28}$$

To reduce the effect of abrupt changes, a sliding window with size N_w is used to smooth the playout rate as [30]

$$\mu(k+1) = \frac{\tilde{\mu}(k+1) + (2^{N_w} - 1) * \mu(k)}{2^{N_w}}. \tag{8.29}$$

Because the video content is usually time varying, the window size should be set properly. The initial value of the window size is 32.

Assume that the ranges of playout rate corresponding to low motion, medium motion and high motion are Ω_l, Ω_m and Ω_h, respectively. It can be shown from equations (8.26) and (8.27) that

$$\Omega_h \subseteq \Omega_m \subseteq \Omega_l. \tag{8.30}$$

When the previous frame is with high (medium) motion and the current frame is with medium (low) motion, a large window size can be used to smooth the

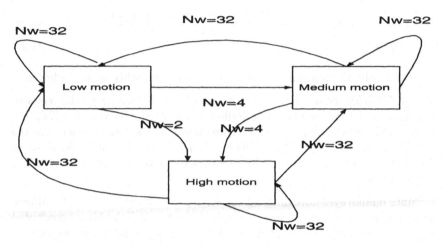

Fig. 8.17. The switching of window size

adjustment. However, when the pervious frame is with medium (low) motion and the current frame is with high (medium) motion, a small window size should be adopted to force the playout rate in the range as soon as possible.

Overall, the window size is switched according to the video contents and is illustrated in Figure 8.17. Three examples are given below to test the performance of our proposed content-based adaptive AMP scheme.

Example 8.4.1. **Slow Time Varying Arrival Rate**

We shall first test our buffer regulation scheme. Assume that the original frame rate is 30f/s. The threshold N_{adapt} is selected as 300. The upper and lower bounds are 38 and 23, respectively, in our AMP scheme. The AMP proposed in [80] is

$$\mu(k) = \begin{cases} 23 & B(k) < 300 \\ 30 & B(k) = 300 \\ 38 & B(k) > 300 \end{cases}. \tag{8.31}$$

In other words, the slowdown factor is $\kappa_1 = 1.3$ and the speed up factor is $\kappa_2 = 0.79$.

The arrival rate is slowly varying piecewise continuous as shown in Fig. 8.18. The buffer occupancy and playout rate are illustrated in Figs 8.19 and 8.20, respectively. It is shown that both our AMP scheme and those proposed by Yuang et al. [214] and Karlman et al. [80] can be used to avoid the buffer underflow. However, the playout rate is adjusted more smoothly by our AMP scheme than those by the existing AMP schemes. □

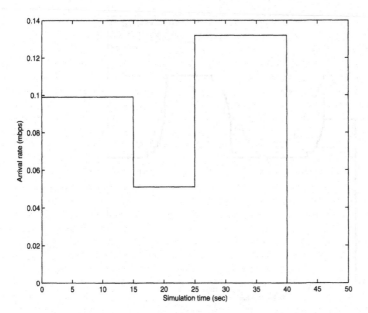

Fig. 8.18. Slowly time varying arrival rate

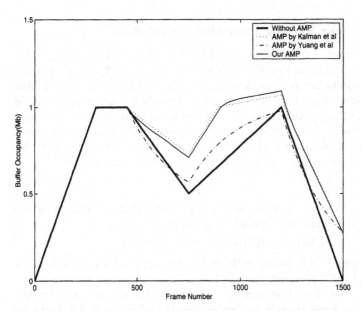

Fig. 8.19. The comparison of buffer occupancy among different schemes

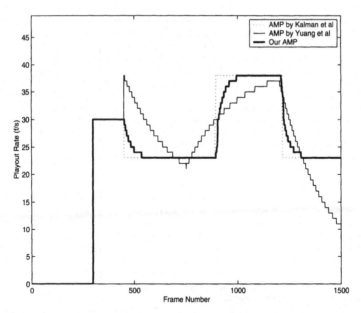

Fig. 8.20. The comparison of playout rate among different AMP schemes

Example 8.4.2. **Fast Varying Piecewise Arrival Rate**

The arrival rate is demonstrated in Fig. 8.21. The buffer occupancy and playout rate are shown in Figs 8.22 and 8.23, respectively. When the playout rate is fixed, the buffer is underflowed after the 1410th frame. The 1410th frame is repeated from then on. Although the AMP scheme proposed by Yuang et al [214] and Kalman et al. [80] can be used to prevent the buffer from underflowing, the playout rate is also changed frequently, the visual quality is sometimes unacceptable. Our scheme can be used to overcome this and the playout rate is also adjusted very smoothly by our scheme.   □

Example 8.4.3. In this example, we test the overall performance of our scheme. The polynomial function $\varphi()$ is given by

$$\varphi(\bar{AH}(t,k), \bar{AE}(t,k)) = \bar{AH}(t,k)\bar{AE}(t,k).$$

The video sequence is Foreman with QCIF size and the frame rate is 30f/s. The bit rate of the coding process is 140kb/s and the arrival pattern is illustrated in Fig. 8.24. The threshold N_{adapt} is selected as 60. The initial values of the upper and lower bounds are selected as 38 and 23, respectively, in our AMP scheme. The AMP proposed in [80] is

$$\mu(k) = \begin{cases} 23 & B(k) < 60 \\ 30 & B(k) = 60 \\ 38 & B(k) > 60 \end{cases}. \tag{8.32}$$

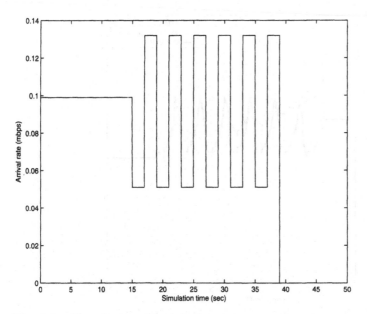

Fig. 8.21. Fast time varying arrival rate

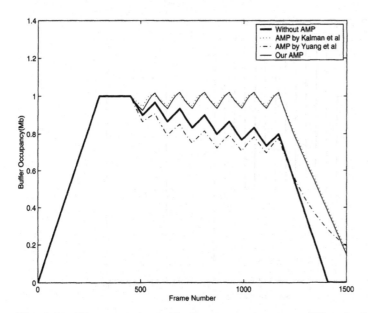

Fig. 8.22. The comparison of buffer occupancy among different schemes

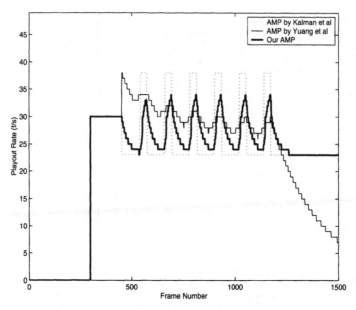

Fig. 8.23. The comparison of playout rate among different AMP schemes

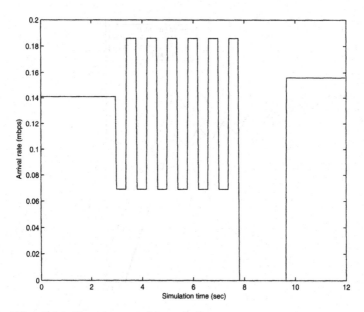

Fig. 8.24. Fast time varying arrival rate

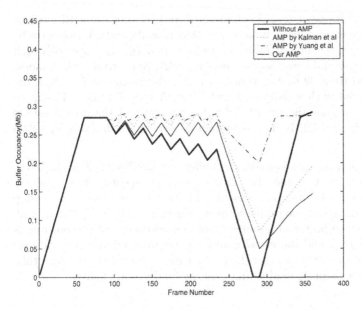

Fig. 8.25. The comparison of buffer occupancy among different schemes

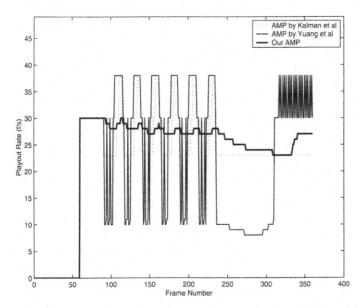

Fig. 8.26. The comparison of playout rate among different AMP schemes

The PME of frames 1 to 273 is between 4 and 16, they are with medium motion, the upper and lower bounds are chosen as 38 and 23, respectively. The PME of the remaining frames is greater than 16, they are with high motion, the upper and lower bounds are 34 and 27, respectively. The window size is switched from 32 to 4 at time $t = 11.1$s. Meanwhile, at $t = 26/3$, the buffer is detected with a danger of underflow. A special request is sent to the sender to increase the priority of video packets and the arrival rate is increased at time $t = 29/3$s. The experimental results are shown in Figs 8.25 and 8.26.

When the playout rate is fixed, the buffer is underflowed after the 281th frame until the 290th frame. The 281th frame is repeated from then on. The motion judder is thus very serious and the visual quality is annoying. Although the AMP scheme proposed by Yuang et al [214] and Kalman et al. [80] can be used to prevent the buffer from underflowing, the playout rate is changed frequently, and the visual quality is still unacceptable. Our scheme can be used to overcome this and the playout rate is adjusted very smoothly. Thus, our scheme can be used to improve the final visual quality.   □

9. Switched Scalable Video Coding Systems

Scalable video coding means that a video sequence is only encoded once and the coded bitstream has the capability of sending any sub-bitstream relevant to a demand/request. In other words, the required sub-bitstream is simply read and packetized from the compressed bitstream. A video coding system that needs to know the scalable-point of the demands prior to the coding process or one that achieves scalability by transcoding is not truly scalable. Certainly, the coding efficiency can be improved by using the priori knowledge of the desired scalable-point in a scalable video coding system.

As introduced in chapter 1, the states of a scalable video coding system are the given bit rate, resolution and frame rate, the control inputs are the motion information and the residual image to be coded. It is very difficult or even impossible to set up a linear model of the form (3.1) or a nonlinear model of the form (4.1) for an SVC system.

However, the switched control can be used to improve the coding efficiency. The whole set of X, Ω, can be divided into n subsets $\Omega_i (i = 1, 2, \cdots, n)$. A switched control can then be designed by

$$\tilde{U}(t) = K_i(X(t)) \quad \text{when} \quad X(t) \in \Omega_i. \tag{9.1}$$

Specifically, the following two major tasks should be completed.

Task 1. Within the Ω_i, the following two items should be determined for each frame.

- The motion information to be coded;

- The residual information to be coded.

This is analogous to the design of each sub-controller in Chapters 3 and 4.

Task 2. The switching point from one group of motion information and residual images to another group should be determined. This is analogous to the switching law of sub-controllers.

9.1 Scalable Video Coding

Reliable transmission of video over heterogeneous networks requires efficient coding, as well as scalability to different client capabilities, system resources, and network conditions. For example, clients may have different display resolutions, systems may have different caching or intermediate storage resources, and networks may have varying bandwidths, loss rates, and best-effort service, relative differentiated QoS or QoS capabilities. Scalable video coding has been proposed to increase its adaptability to network and client conditions. There are many applications which require scalable and reliable video coding. One typical example is given below [143].

Example 9.1.1. **Video Streaming over Heterogeneous IP Networks**

With the rapid development of IP networks, both wired and wireless, more and more video contents are streamed to the end user either live or on-demand. There are many challenges in streaming video over the IP networks.

The IP networks are heterogeneous networks, the connections between server and clients are different from one user to the other. One user may choose the best effort service, the other one may select the relative differentiated QoS, and yet another may use the QoS. When the users choose the best effort service or the relative differentiated QoS, the available bandwidth can change vastly. The connection bandwidth varies from 9.6kbps to 100Mbps and above.

Further, the user devices have different resolutions and different computation power. Different users would have different requirements of the playout rate, and so on.

Unfortunately, most of the current streaming services are provided through non-scalable coding technologies. There are many issues to be resolved when dealing with the above challenges. For example, the service providers have to encode the same content into a very large number of bitstreams to support different network connections and device types. If the number is limited, many users would not achieve the potential high quality they could get. Moreover, the nonscalable coding technologies prevent fast switching between different bit rates and they diminish the adaptivity and prevent recovery from errors or packet losses gracefully, even with switched P-frame (SP) scheme in the advance video code (AVC).

Scalable video coding (fine-granularity) can resolve many of the above issues.

It may be very rare for more than one users to be interested in the same program stored at a remote server at the same time. However, SVC is still useful, especially in the case where a local server can be used to temporarily store the program. An example is illustrated in Fig. 9.1. Assume that a new digital program is produced and stored at a remote server. User C is the first

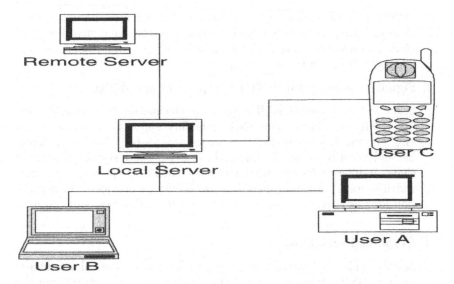

Fig. 9.1. A situation where SVC is useful

one to watch the program with the frame rate of 15f/s, resolution as CIF and the bit rate at 128kb/s. The bitstream can be downloaded from the remote server and a copy is stored at a local server. User A is the second one to watch the program with the frame rate af 30f/s, resolution as 4CIF and the bit rate at 512kb/s. With the SVC, only the enhancement layers need to be downloaded from the remote server. Similarly, a copy can be stored at the local server. User B is the last one to watch the program with the frame rate of 30f/s, resolution as 4CIF and the bit rate at 1024kb/s. In this situation, only the further enhancement layers need to be downloaded from the remote server. The stored bitstream will be destroyed after certain period.   □

There are many types of scalabilities [143]. Four of them are listed as follows:

1. **Spatial scalability**

   Scalable video coding shall support a mechanism that enables spatial scalability. The spatial scalability should be able to support a variety of resolutions, including common intermediate format (CIF, 352×288), QCIF(176×144), 4CIF(704×576), and also higher resolutions such as 1436 × 1080 to 3610 × 1536. For instance, for multi-channel content production and distribution, the same stream will be viewed on a variety of devices having different spatial resolutions.

2. **Temporal scalability**

   Scalable video coding shall support a mechanism that enables temporal scalability. The temporal scalability supports decoding of moving pictures with frame rates ranging between 5Hz and 75 Hz. For instance, for multi-channel content production and distribution, the same stream will be viewed on a variety of devices having different temporal resolutions, like 7.5Hz, 15Hz and 30Hz, and so on.

3. **Signal to noise ratio (SNR) (or quality) scalability**

Scalable video coding shall support a mechanism that enables quality (SNR) scalability. The SNR scalability supports decoding of moving pictures having quality that varies progressively between acceptable and visually lossless in fine-grained steps. For instance, in order to charge a different fee for higher resolution content (requiring more bandwidth/storage), different quality levels should be provided. An application for this scalability could be in the context of storage and transmission.

4. **Combined scalability**

Scalable video coding shall support a mechanism that enables combined quality (SNR), temporal and spatial scalability. For instance, when a device moves from a high bandwidth to a low bandwidth connection, the change in the user experience should be gradual. An example is given below.

Example 9.1.2. **Combined SNR and Spatial Scalability**

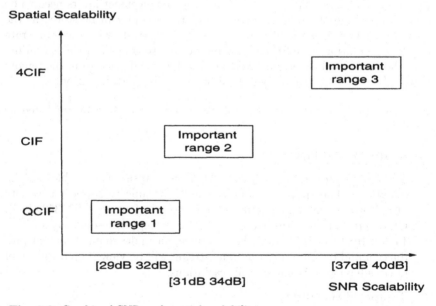

Fig. 9.2. Combined SNR and spatial scalability

A typical example with combined SNR and spatial scalability is illustrated in Fig. 9.2. The important ranges are determined by the possible practical applications.   □

9.2 Conventional Scalable Video Coding

9.2.1 Motion Estimation/Motion Compensation (ME/MC)

In a typical scene, there will be a great deal of similarity between nearby frames of the same video sequence. If a stripe of motion picture film is examined, it will be found that the repetitive nature of the frames is evident. Many video sequences consist of a static background with one or more moving foreground objects. The temporal redundancy between the static background in two adjacent frames can be removed by simply predicting the background in the current frame from that of the previous frame at the same place. However, the moving foreground objects of the two adjacent frames are not in the same place. If we can measure the displacement, we can still generate a residual image by using the original image and the reference image. The obtained residual image contains less energy than the original one, and thus there are fewer bits to code. Obviously, we must send the decoder the motion vectors (MVs) to inform the decoder exactly where in the previous frame to get the object. The MV is a two-dimensional value (dx, dy), normally represented by a horizontal component dx and a vertical component dy. The process of obtaining the MV is known as *motion estimation*, while using the MV to reduce or eliminate the effects of motion is known as *motion compensation*. Unfortunately, the transmission overhead needed to inform the decoder of the true motion at every pixel in the frame may far out-weight the advantages of MC. Thus, all widely known video compression algorithms force the decoder to operate with only limited information about the motion in the scene. The most popular method for the ME/MC is based on block matching which treats each block in the current frame separately and searches for an identical or similar block in the reference frame. This MV (dx, dy) is often selected by the encoder to minimize the following performance index:

$$J = D_p(dx, dy) + \lambda_{opt}(R_{mv}(dx, dy) + R_{ref}(F_2),$$ (9.2)

where $R_{mv}(dx, dy)$ and $R_{ref}(F_2)$ are the number of bits used to code the MV (dx, dy), and the number of bits to code the reference frame F_2, respectively, and $D_p(dx, dy)$ is defined as

$$D_p(dx, dy) = \sum_{x,y} |F_1(x, y) - F_2(x - dx, y - dy)|$$ (9.3)

with $F_1(x, y)$ and $F_2(x, y)$ be the predicted frame and the reference frame, respectively.

In other words, the objective of ME/MC is to remove redundancies among frames by prediction. The overall prediction depends on the number of referencing frames, the accuracy level of MVs, the choice of partition modes of a MB, the distance between the predicted frame and the referencing frames, and the quality of referencing frames. The prediction gain increases with an increase in the number of referencing frames, the accuracy of MVs, the choice of partition modes, or the quality of referencing frames, while it decreases with an increase in the distance between the predicted frame and the referencing frames.

There are two types of ME/MC schemes which are determined by the reference frame. It is a closed-loop based scheme if the reference frame is a reconstructed one. The closed-loop based ME/MC is used in the existing video coding standards. The early ME/MC technique adopted in MPEG-1 is extremely simple in the sense that the block size is fixed as 16×16 and MVs are only at integer-pixel level [73]. The number of reference frames for a P frame is 1, and that for a B frame is 2. To improve the prediction gain, MPEG-2 extends the accuracy of MVs to half-pixel level [74], while the number of referencing frames is unchanged. MPEG-4 not only further extends the accuracy to quarter pixel level but also allows 8×8 block as a unit of ME/MC [72]. However, the number of referencing frames is not changed. To more accurately compensate local motion among pictures, the latest video coding standard H.264/AVC adopts the adaptive block-size ME/MC technique [192]. The unit for ME/MC can be selected from one of the seven modes: 16×16, 16×8, 8×16, 8×8, 4×8, 8×4 and 4×4. Meanwhile, the number of referencing frame for a P frame can be greater than 1, and that for B frame can be greater than 2.

If the reference frame is the original one, it is then an open-loop based scheme. The open-loop based ME/MC is called motion compensation temporal filtering (MCTF). The MCTF was proposed by Ohm [142] as an efficient tool to remove temporal redundancies in wavelet based video coding schemes. Choi and Woods [33] improved it by making the direction of the ME the same as that of MC. Hierarchical variable size block matching (HVSBM) is a popular method to search for the motion vector field (MVF) in the MCTF. The HVSBM consists of constructing an initial full MV tree and pruning it with a given Lagrangian multiplier λ_{opt}. The initial tree is generated by combining the splitting operation with hierarchical ME/MC. At each level, MVs of all internal nodes as well as the leaf nodes are refined. Each leaf node is split into child blocks and their MVs are estimated. A node is split only if it decreases the total motion estimation error. Since the MV (dx, dy) is often selected by the encoder to minimize the performance index (9.2), the initial full MV tree is pruned with the Lagrangian multiplier λ_{opt} and the final block partition mode is achieved.

9.2.2 Hybrid Video Coding

Because of a high similarity between adjacent frames, an efficient video coding technique performs an effective removal on temporal redundancy. The traditional video compression method is based on hybrid coding [59], which employs ME/MC in the temporal domain and two-dimensional discrete cosine transform (DCT) algorithm in the spatial domain.

Based on the group of pictures (GOP) structure, the encoder side needs to distinguish them from among three kinds of frames, I, P and B. Each I-frame is then divided into blocks without performing any ME/MC. However, the INTRA prediction can be used to remove the spatial redundancy [192]. In an I-frame, each macroblock (MB) is coded as follows: Each 8x8 (or 4x4) block of pixels in an MB undergoes a DCT transform to form an 8x8 (or 4x4) array of transform coefficients. The transform coefficients are then quantized with a variable quantizer matrix. Quantization involves dividing each DCT coefficient by a quantization step size. The resulting quantized DCT coefficients are scanned (e.g. using Zig-Zag scanning) to form a sequence of DCT coefficients. The DCT coefficients are then organized into run-level pairs. Finally, the run-level pairs are encoded via a variable length code [192].

In a P frame, a decision is made to either code each MB as an I MB, which MB is then encoded will be done according to the technique described above, or code the MB as a P MB. For each P MB, a prediction of the MB in the previous P (or I) frame is obtained. The resulted residual between the predicted MB and the current MB is then coded via the DCT, quantization, Zig-Zag scanning, run-level pair encoding and VLC coding. This is illustrated in Fig. 9.3.

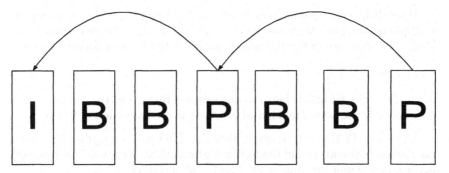

Fig. 9.3. Backward ME/MC for P frames

In the coding of a B-frame, a decision has to be made on the coding of each MB. The choices are (1) an I MB, (2) a unidirectional forward predictive MB, (3) a unidirectional backward predictive MB, and (4) a bidirectional predictive MB. In the case of forward, backward and bidirectional ME/MC predictions, the residual is encoded using the DCT, quantization, Zig-Zag scanning, run-level pair encoding and VLC coding. An example is shown in Fig. 9.4.

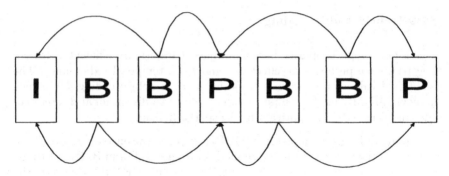

Fig. 9.4. Bi-directional ME/MC for B frames

B frames have the smallest number of bits when encoded, then P frames, with I frames having the most number of bits. Thus, the greatest degree of compression is achieved for B frames. For each of the I, B and P frames, the number of bits resulting from the encoding process can be controlled by setting the quatization step size of each MB [31].

9.2.3 Rate Distortion Optimization

Another key problem in video compression is the operation control of the source encoder. The task of coder control is to determine a set of coding parameters, like transform coefficients, quantization parameters, MVs and so on, such that a certain rate-distortion tradeoff is achieved for a given decoder [193].

Rate distortion optimization (RDO) with the utilization of Lagrangian multipliers (λ_{opt}) has been adopted in H.264 as an efficient coder control tool [192]. The objective of the RDO is to minimize the joint performance index

$$J = D + \lambda_{opt}^2 \chi, \tag{9.4}$$

where D is the distortion between the original frame and the reconstructed one, and χ is the actual number of bits that is used to code the frame.

Equations (9.2) and (9.4) imply that the RDO is also a trade-off between the motion information and the residual information. Obviously, the trade-off plays a very important role in the hybrid video coding.

Each solution to minimize J in (9.4) for a given value of the Lagrangian multiplier λ_{opt} corresponds to an optimal solution to the following constrained optimization problem for a particular value of R_c [181]:

$$\min\{D\}, \quad \text{subject to } \chi < R_c. \tag{9.5}$$

Let QP denote the quantization parameter for the coding process. It can be derived that [192]

$$\lambda_{opt} = 0.92 * 2^{(QP-12)/6}. \tag{9.6}$$

Obviously, there is a one-to-one map between λ_{opt} and QP.

For a given quantization parameter QP, the Lagrangian multiplier λ_{opt} is computed by equation (9.6). For each possible MB partition of the current MB, the reference frame and the associated MVs are determined below [165].

1. For all sub-MB j of the ith MB in the kth frame, find the best matching sub-MB in the lth reference frame by minimizing J in (9.2). Suppose that the corresponding MV is denoted by (dx, dy).

2. For the given MVs, find the reference frame for the ith MB in the kth frame as

$$l(k,i) = \text{argmin}_l\{\sum_j (D_p(dx, dy) + \lambda_{opt}(R_{mv}(dx, dy) + R_{ref}(l))\}. \tag{9.7}$$

With the reference frame $l(k,i)$, the overall cost is

$$J(k,i) = \sum_j (D_p(dx, dy) + \lambda_{opt}(R_{mv}(dx, dy) + R_{ref}(l(k,i))). \tag{9.8}$$

3. Given the reference frames and the associated MVs of the forward ME/MC and the backward ME/MC, the overall cost of a bi-direction ME/MC can be computed by using an iterative method [165]. The final ME/MC is chosen as that has the minimal cost.

An example is now provided to show the significance of computing λ_{opt} by equation (9.6).

Example 9.2.1. Consider the video sequence Foreman with QCIF size and suppose that the frame rate is 15f/s. The experimental results are listed in Table 9.1 where the first four rows contain the results with λ_{opt} computed by equation (9.6) while the last two rows contain the results with any other λ_{opt} not computed by equation (9.6). Clearly, the coding efficiency is dropped significantly if λ_{opt} is not computed by equation (9.6).   □

Table 9.1. The choices of λ_{opt} for the RDO

(QP, λ_{opt})	Y (PSNR, dB)	Bit Rate (kb/s)
(36,14.6)	30.63	31.45
(29,6.6)	35.02	79.7
(28,5.9)	35.72	92.92
(26, 4.6)	37.06	122.96
(36,4.6)	33.44	78.88
(26,14.6)	32.03	90.19

9.2.4 Mutual Referencing Frame Pair

In Chapters 2-4, we illustrated that a bigger unit, i.e. a cycle instead of a CVS, can be used to derive less conservative conditions on the stability of switched and impulsive systems. In this section, we shall demonstrate that a bigger unit, i.e. mutual referencing frame pair, can also be applied to improve the coding efficiency.

A mutual referencing frame pair is defined below.

Definition 9.2.1. *A pair of mutual referencing frames is composed of two frames F_2 and F_3, where F_2 is a reference frame of F_3, F_3 is also a reference frame of F_2, and F_2 and F_3 are decodable at the decoder side.*   □

The concept is illustrated in Fig. 9.5 and it is very useful for the MCTF with non-dyadic structures [184] and the IBBPBBP in the H.264 [192]. Note that a special case of mutual referencing frame pair was studied in [184].

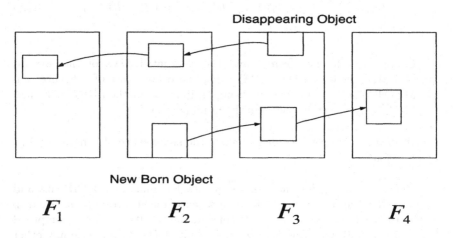

Disappearing Object

New Born Object

$$F_1 \qquad\qquad F_2 \qquad\qquad F_3 \qquad\qquad F_4$$

Fig. 9.5. A pair of mutual referencing frames

The mutual referencing frame pair is not allowed in the existing video coding standards. However, the coding efficiency can be improved with it. Before providing two examples to illustrate this, the concept of cycle presented in Chapters 2-4 is extended to the video coding.

Let F_k be a frame and an arrow be a ME/MC in Fig. 9.6. A logical path is a sequence of frames where a frame in the sequence is a reference frame of its subsequent one. For example, $F_1\ F_2\ F_3\ F_5\ F_6\ F_7\ F_3$ is a logical path. The length of a logical path is the total number of frames in the logical path. For example, the length of $F_1\ F_2\ F_3\ F_5\ F_6\ F_7\ F_3$ is 7. Frame F_l connects to F_k if there is a logical path which starts from F_k and ends at F_l. For example, F_7

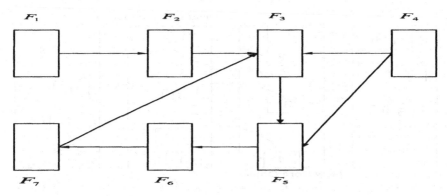

Fig. 9.6. Illustration of concepts in graph theory

connects to F_1. A logical path is closed if the first frame in the path is the same as the last one in the path. A closed path in which no frame appears more than once except for the one that is the first and the last is a cycle. For example, F_3 F_5 F_6 F_7 F_3 is a cycle. They can also be defined at the block level.

We shall also introduce a concept of minimal motion compensation unit (MMCU) which is defined as the minimal unit for the ME/MC. For example, 4 by 4 is the MMCU for the H.264 [192].

With the concepts of cycle and MMCU, two examples are given below.

Example 9.2.2. **No Cycle at the MMCU Level**

Consider the example given in Fig. 9.5. The referencing block for block 1 of frame F_2 is block 1 of frame F_3. Meanwhile, the referencing block for block 2 of frame B3 is block 2 of frame F_2. Obviously, frames F_2 and F_3 are a mutual referencing frame pair. There does not exist any cycle at the MMCU level.

This example is applicable to both the closed-loop and the open-loop based ME/MC. Thus, the concept of mutual referencing frame pair can be applied to improve the coding efficiency of B frames when there is more than one B frames between two successive P frames. □

Example 9.2.3. **Several Cycles at the MMCU Level**

An example is provided in Fig. 9.7. Frames F_2 and F_3 are a mutual referencing frame pair. There are three cycles at the MMCU level.

The MC among MMCUs at the encoder side is

$$\begin{bmatrix} L_1(3) \\ H_1(3) \\ H_2(2) \\ L_2(1) \end{bmatrix} = \begin{bmatrix} 1 & 0 & 0 & 0 \\ -0.5 & 1 & -0.5 & 0 \\ 0 & -0.5 & 1 & -0.5 \\ -1 & 0 & 0 & 1 \end{bmatrix} \begin{bmatrix} F_1(3) \\ F_2(3) \\ F_3(2) \\ F_4(1) \end{bmatrix}, \tag{9.9}$$

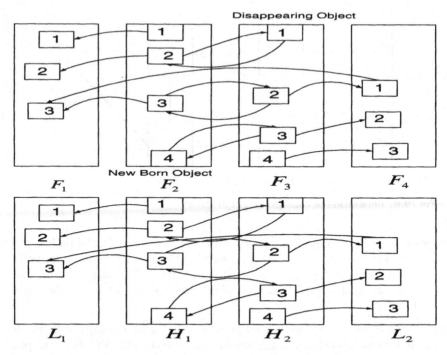

Fig. 9.7. A mutual referencing frame pair with cycle at the MMCU level

$$\begin{bmatrix} H_1(4) \\ H_2(3) \\ L_2(2) \end{bmatrix} = \begin{bmatrix} 1 & -1 & 0 \\ -0.5 & 1 & -0.5 \\ 0 & 0 & 1 \end{bmatrix} \begin{bmatrix} F_2(4) \\ F_3(3) \\ F_4(2) \end{bmatrix}, \tag{9.10}$$

$$\begin{bmatrix} L_1(2) \\ H_1(2) \\ H_2(1) \end{bmatrix} = \begin{bmatrix} 1 & 0 & 0 \\ -0.5 & 1 & -0.5 \\ 0 & -1 & 1 \end{bmatrix} \begin{bmatrix} F_1(2) \\ F_2(2) \\ F_3(1) \end{bmatrix}. \tag{9.11}$$

The corresponding MC among MMCUs at the decoder side is

$$\begin{bmatrix} F_1(3) \\ F_2(3) \\ F_3(2) \\ F_4(1) \end{bmatrix} = \begin{bmatrix} 1 & 0 & 0 & 0 \\ 1 & 4/3 & 2/3 & 1/3 \\ 1 & 2/3 & 4/3 & 2/3 \\ 1 & 0 & 0 & 1 \end{bmatrix} \begin{bmatrix} L_1(3) \\ H_1(3) \\ H_2(2) \\ L_2(1) \end{bmatrix}, \tag{9.12}$$

$$\begin{bmatrix} F_2(4) \\ F_3(3) \\ F_4(2) \end{bmatrix} = \begin{bmatrix} 2 & 2 & 1 \\ 1 & 2 & 1 \\ 0 & 0 & 1 \end{bmatrix} \begin{bmatrix} H_1(4) \\ H_2(3) \\ L_2(2) \end{bmatrix}, \tag{9.13}$$

$$\begin{bmatrix} F_1(2) \\ F_2(2) \\ F_3(1) \end{bmatrix} = \begin{bmatrix} 1 & 0 & 0 \\ 1 & 2 & 1 \\ 1 & 2 & 2 \end{bmatrix} \begin{bmatrix} L_1(2) \\ H_1(2) \\ H_2(1) \end{bmatrix}. \tag{9.14}$$

This example is only applicable to the open-loop based ME/MC.   □

The concept of motion threading was introduced by Xu et al. in [199]. Each thread starts with an emanating block and ends with a terminating block. Blocks (if any) in the middle of a thread are continuing blocks. The objective of motion threading is to form as many long thread as possible because increasing short threads will significantly increase the number of artificial boundaries and degrade the coding performance. Here, the concept of motion threading and the concept of mutual referencing frame pair are used together to improve the coding efficiency. The motion threading is thus restricted among two L frames and H frames between them, or two P frames and B frames between them, or one I frame and one P frame and B frames between them. For example, in Fig. 9.7, block $F_1(3)$ is an emanating block, blocks $F_2(3)$ and $F_3(2)$ are continuing blocks, and block $F_4(1)$ is a terminating block. The L frames (I frame, or P frames) are called the boundaries of a motion threading. The number of H frames (or B frames) is determined by the computational complexity and the available memory size for the coding process.

Example 9.2.4. Consider Fig. 9.8 where there are four high subbands between two low subbands. The MC among MMCUs at the encoder side is

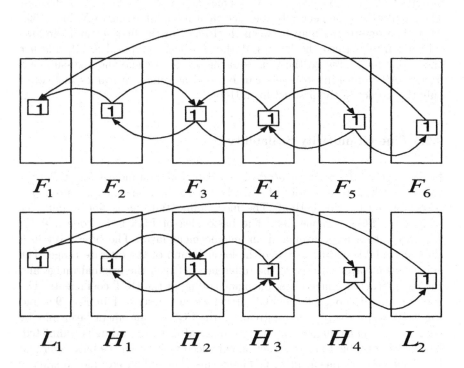

Fig. 9.8. Four high subbands between two low subbands

$$
\begin{bmatrix} L_1(1) \\ H_1(1) \\ H_2(1) \\ H_3(1) \\ H_4(1) \\ L_2(1) \end{bmatrix} = \begin{bmatrix} 1 & 0 & 0 & 0 & 0 & 0 \\ -0.5 & 1 & -0.5 & 0 & 0 & 0 \\ 0 & -0.5 & 1 & -0.5 & 0 & 0 \\ 0 & 0 & -0.5 & 1 & -0.5 & 0 \\ 0 & 0 & 0 & -0.5 & 1 & -0.5 \\ -1 & 0 & 0 & 0 & 0 & 1 \end{bmatrix} \begin{bmatrix} F_1(1) \\ F_2(1) \\ F_3(1) \\ F_4(1) \\ F_5(1) \\ F_6(1) \end{bmatrix}.
$$

The corresponding MC among MMCUs at the decoder side is

$$
\begin{bmatrix} F_1(1) \\ F_2(1) \\ F_3(1) \\ F_4(1) \\ F_5(1) \\ F_6(1) \end{bmatrix} = \begin{bmatrix} 1 & 0 & 0 & 0 & 0 & 0 \\ 1 & 1.6 & 1.2 & 0.8 & 0.4 & 0.2 \\ 1 & 1.2 & 2.4 & 1.6 & 0.8 & 0.4 \\ 1 & 0.8 & 1.6 & 2.4 & 1.2 & 0.6 \\ 1 & 0.4 & 0.8 & 1.2 & 1.6 & 0.8 \\ 1 & 0 & 0 & 0 & 0 & 1 \end{bmatrix} \begin{bmatrix} L_1(1) \\ H_1(1) \\ H_2(1) \\ H_3(1) \\ H_4(1) \\ L_2(1) \end{bmatrix}.
$$

$\square$

Example 9.2.5. **Frame Partition at a Temporal Level in SVC**

Suppose that we have a total of N_l frames at the lth temporal level. To generate the frames at the $(l+1)$th temporal level, the frames are partitioned into a set of N_{l+1} $(0 < N_{l+1} < N_l)$ frames and a set of $(N_l - N_{l+1})$ frames. The partition can be generally specified by using a bit string of N_l bits [166]. There is a one-to-one map between the frames in the first set and the low-subband frames (also the frames at the $(l + 1)$th temporal level), another one-to-one map between the frames in the second set and the high-subband frames. All frames in both sets can be used as reference frames instead of only the frames in the first set as in [166]. $\square$

9.2.5 Fine Granularity Scalability

In response to the growing need on a video coding standard for streaming video over the Internet, a lot of scalable video coding schemes are developed based on the hybrid method. The most influential one is fine granularity scalability (FGS) scheme [93]. The basic idea of FGS is to code a video sequence into a base layer and an enhancement layer. The base layer uses the non-scalable coding to reach the lower bound of the bit-rate range. The enhancement layer is to code the difference between the original image and the reconstructed image using bit-plane coding of the DCT coefficients. The encoder and decoder of the FGS structure are described in Fig. 9.9 and Fig. 9.10, respectively. The bitstream of the FGS enhancement layer may be truncated into any number of bits per image after the encoding is completed. The decoder is able to reconstruct an enhanced video from the base layer and the truncated enhancement layer bitstreams. The enhanced video quality is proportional to the number of bits decoded for each picture. This hybrid scalable video coding scheme has a number of advantages, such as the use of mature technology in the DCT algorithm, easy to provide low delay solution,

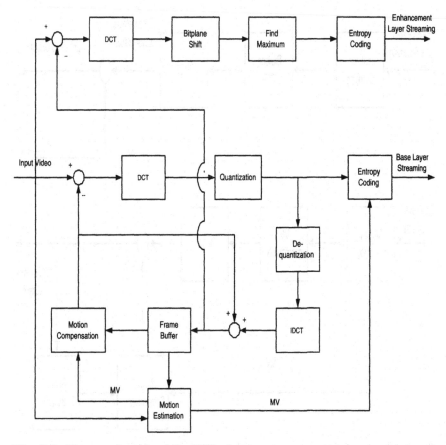

Fig. 9.9. The encoder side of the FGS scheme

and easy to implement with the existing video coding standards. On the other hand, although there are some improvements for FGS, such as in [159, 158, 197, 67], there are still some disadvantages in this method. Two of them are listed below.

1. In order to avoid error propagation, the FGS uses only the base layer frames as reference frames for temporal prediction. In this case, the prediction gain and consequently the coding efficiency will be reduced, especially at high bitrate. Although part of the enhancement layer is involved in the prediction in the enhanced FGS [197], the coding efficiency is still very low. One of the reasons is as follows:

   It is well known that the trade-off between the motion information and the residual image is most crucial for a scalable video coding (SVC) scheme. Since the motion information is generated at low bit rate in the FGS, and

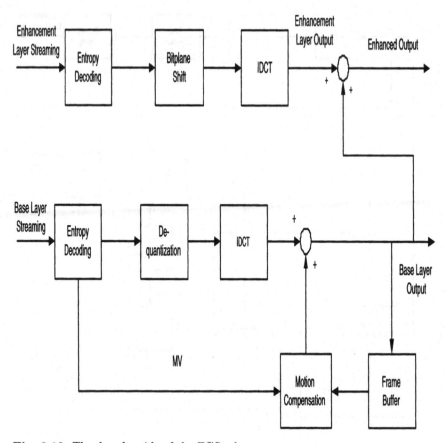

Fig. 9.10. The decoder side of the FGS scheme

it is not enough at high bit rate, the coding efficiency is still low even with the enhanced FGS schemes [197, 67].

2. The hybrid coding scheme employs ME/MC at a fixed resolution frame, so it is difficult to achieve spatial scalability, especially hybrid scalability.

9.3 Three Dimensional Subband Wavelet Coding

9.3.1 Motion Compensated Temporal Filtering

Three dimensional (3D) subband wavelet coding was proposed as an efficient SVC scheme, especially with the introduction of the MCTF [142, 33]. The subband video coding, in general, refers to the class of video coding techniques where by parallel application of a set of filters the input video is decomposed

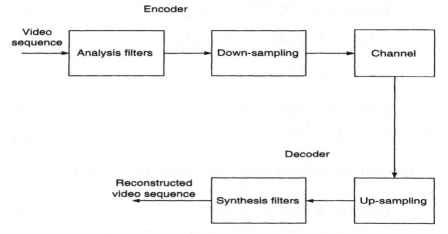

Fig. 9.11. The analysis/synthesis system

into several narrow bands. The resulting bands are then decimated and coded separately for the purpose of transmission. For reconstruction, decimated video signals are decoded, interpolated, and filtered before being added to reproduce the original video sequence. This is illustrated in Fig. 9.11.

The existing MCTF structure can be divided into two classes: dyadic multi-resolution schemes and non-dyadic structures. For the sake of simplicity, we shall present the equations for the MCTF without taking the motion aspect into account.

1. **Dyadic structures**

   Let the original frames, the high subbands, and the low subbands be denoted by F_k, H_k and L_k respectively. The first MCTF scheme is the HAAR one [142] and the analysis is

$$\text{Prediction: } H_k = F_{2k+1} - F_{2k}, \tag{9.15}$$

$$\text{Update: } L_k = \nu_1 F_{2k} + \frac{\nu_2}{2} H_k. \tag{9.16}$$

   where ν_1 and ν_2 are two parameters.

   The corresponding synthesis is

$$F_{2k} = L_k - \frac{1}{2} H_k, \tag{9.17}$$

$$F_{2k+1} = L_k + \frac{1}{2} H_k. \tag{9.18}$$

   The 5/3 MCTF is the next popular dyadic structure and the analysis is

$$\text{Prediction:} \quad H_k = F_{2k+1} - \frac{1}{2}(F_{2k} + F_{2k+2}), \tag{9.19}$$

$$\text{Update:} \quad L_k = \nu_1 F_{2k} + \frac{\nu_2}{4}(H_{k-1} + H_k). \tag{9.20}$$

The corresponding synthesis is

$$F_{2k} = (L_k - \frac{\nu_2}{4}(H_{k-1} + H_k))/\nu_1, \tag{9.21}$$

$$F_{2k+1} = H_k + \frac{1}{2}(F_{2k} + F_{2k+2}). \tag{9.22}$$

Generally, the MCTF of dyadic structure can be represented by

$$\text{Prediction:} \quad H_k = F_{2k+1} - \sum_{i=\varrho_1}^{\varrho_2} a_i F_{2k-2i}, \tag{9.23}$$

$$\text{Update:} \quad L_k = \nu_1 F_{2k} + \nu_2 \sum_{i=\varrho_3}^{\varrho_4} c_i H_{k-i}. \tag{9.24}$$

where ϱ_1, ϱ_2, ϱ_3, ϱ_4, a_i and c_i are constants.

For a normalization of low subbands and high subbands, appropriately chosen scaling factors are applied. In practice, these scaling factors do not need to be applied during the decomposition and reconstruction process, but can be incorporated when selecting the quantization step sizes during encoding.

2. **Non-dyadic structures**

The non-dyadic structure used for the SVC is the 3-band of the dyadic Haar MCTF. It has two detailed subbands and a bidirectional update step as follows [183]:

$$\text{Prediction:} \quad H_k^- = F_{3k+1} - F_{3k}, \tag{9.25}$$

$$H_k^+ = F_{3k-1} - F_{3k}, \tag{9.26}$$

$$\text{Update:} \quad L_k = F_{3k} + \frac{1}{4}(H_k^- + H_k^+). \tag{9.27}$$

The corresponding synthesis is

$$F_{3k} = L_k - \frac{1}{4}(H_k^- + H_k^+), \tag{9.28}$$

$$F_{3k-1} = H_k^+ + F_{3k}, \tag{9.29}$$

$$F_{3k+1} = H_k^- + F_{3k}. \tag{9.30}$$

The above MCTF is called an unconstrained MCTF (UMCTF) if the update step is replaced by $L_k = F_{2k}$ (or $L_k = F_{3k}$) [185]. The results that will be provided in the remaining part of this chapter are more suitable for the UMCTF. To help readers understand better the development of our scheme, we shall focus on the MCTF.

9.3.2 The Error Propagation Pattern of the MCTF

The MCTF is usually performed using the 2-step lifting scheme. The first step (prediction step) removes as much energy as possible in the difference frame, while the second step (update step) consists of updating the original frame by re-adding the difference frames generated in the prediction step. The main purpose of the update step is to make the quantization noises of the different frames orthogonal. This enables an efficient compression of the video in the framework of the MCTF and distributes the compression error more uniformly among the compressed frames. Moreover it reduces the temporal aliasing in the low-pass frame, increasing the coding efficiency in some cases.

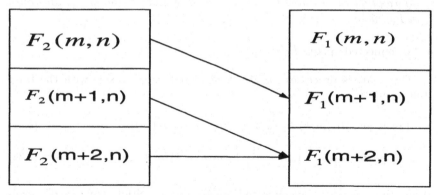

Fig. 9.12. A simple example for the illustration of MCTF

We shall now use an example based on the Haar MCTF to illustrate the analysis and synthesis processes associated with the MCTF. For simplicity, we consider six pixels of two frames F_1 and F_2: $F_1(m, n)$, $F_1(m+1, n)$, $F_1(m+2, n)$, $F_2(m, n)$, $F_2(m+1, n)$ and $F_2(m+2, n)$. The predictions among them are illustrated in Fig. 9.12. They are classified into the following three categories:

- Pair of "single connected" pixels: $F_2(m, n)$ and $F_1(m+1, n)$.

- Group of "multiple connected" pixels: $F_2(m+1, n)$, $F_2(m+2, n)$ and $F_1(m+2, n)$, where $F_2(m+1, n)$ is a better match pixel for $F_1(m+2, n)$.

Remark 9.3.1. It is very important to find a better match pixel for a multiple connected pixel. The block artifacts can be greatly reduced. The motion

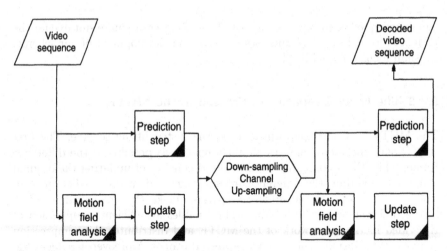

Fig. 9.13. An example to find the best match

information can be used to design a scheme and an example is illustrated in Fig. 9.13.   □

- "unconnected" pixel: $F_1(m, n)$.

The analysis processes for the "single connected" pixels with the Haar transform are

$$\text{Prediction:} \quad H_1(m,n) = \frac{1}{\sqrt{2}}(F_2(m,n) - F_1(m+1,n)), \tag{9.31}$$

$$\text{Update:} \quad L_1(m+1,n) = \frac{1}{\sqrt{2}}(F_2(m,n) + F_1(m+1,n)). \tag{9.32}$$

While those for the "unconnected" pixels and "multiple connected" pixels are

$$\text{Prediction:} \quad H_1(m+1,n) = \frac{1}{\sqrt{2}}(F_2(m+1,n) - F_1(m+2,n)), \tag{9.33}$$

$$H_1(m+2,n) = \frac{1}{\sqrt{2}}(F_2(m+2,n) - F_1(m+2,n)), \tag{9.34}$$

$$\text{Update:} \quad L_1(m,n) = \sqrt{2}F_1(m,n), \tag{9.35}$$

$$L_1(m+2,n) = \frac{1}{\sqrt{2}}(F_2(m+1,n) + F_1(m+2,n)), \tag{9.36}$$

where $L_1(m,n)$ and $H_1(m,n)$ are temporal low and high subbands, $F_1(m,n)$ and $F_2(m,n)$ are the reference and predicted frames.

It can be shown from equation (9.36) that the difference between $F_2(m+1,n)$ and $F_2(m+2,n)$ is that $F_1(m+1,n)$ is involved in the update step. Because of this, different error propagation patterns occur.

The synthesis processes for them are

$$F_1(m+1,n) = \frac{1}{\sqrt{2}}(L_1(m+1,n) - H_1(m,n)),$$

$$F_2(m,n) = \frac{1}{\sqrt{2}}(H_1(m,n) + L_1(m+1,n)),$$

$$F_1(m,n) = \frac{1}{\sqrt{2}}L_1(m,n),$$

$$F_1(m+2,n) = \frac{1}{\sqrt{2}}(L_1(m+2,n) - H_1(m+1,n)),$$

$$F_2(m+1,n) = \frac{1}{\sqrt{2}}(L_1(m+2,n) + H_1(m+1,n)),$$

$$F_2(m+2,n) = \sqrt{2}H_1(m+2,n) + F_1(m+2,n).$$

Let us study the error propagation pattern of the MCTF. Suppose that uncorrected noise terms σ_L^2 and σ_H^2 are applied to L_1 and H_1, respectively. The resulting noise terms for "single connected" pixels are

$$\sigma_{F_1(m+1,n)}^2 = \frac{1}{2}(\sigma_L^2 + \sigma_H^2), \tag{9.37}$$

$$\sigma_{F_2(m,n)}^2 = \frac{1}{2}(\sigma_L^2 + \sigma_H^2). \tag{9.38}$$

It can be shown from the above equations that the prediction step accumulates the noise while the update step distributes the noise over two frames. With both the prediction and update steps, the same amount of noise after the synthesis is transformed from the uncorrected quantization noise to the "single connected" pixels of the reconstructed frames F_1 and F_2 via orthogonal transforms.

However, the noise terms for "unconnected" pixels and "multiple connected" pixels are

$$\sigma_{F_1(m,n)}^2 = \frac{1}{2}\sigma_L^2, \tag{9.39}$$

$$\sigma_{F_1(m+2,n)}^2 = \frac{1}{2}(\sigma_L^2 + \sigma_H^2), \tag{9.40}$$

$$\sigma_{F_2(m+1,n)}^2 = \frac{1}{2}(\sigma_L^2 + \sigma_H^2), \tag{9.41}$$

$$\sigma_{F_2(m+2,n)}^2 = \frac{\varepsilon}{2}\sigma_H^2 + \sigma_{F_1(m+2,n)}^2, \tag{9.42}$$

where ε is a constant dependent on video sequences.

It can be seen from equations (9.41) and (9.42) that different error propagation pattern occurs on $F_2(m+1,n)$ and $F_2(m+2,n)$. Moreover, it can be shown from equations (9.37)-(9.42) that, in any one ME/MC pair, the higher the number of unconnected pixels in the to-be-lowpass-filtered frame, the lower is its quantization noise in the temporal low frequency frame, and hence the better its reconstructed PSNR. This better reconstructed PSNR is only for temporal low frequency frames, and the overall PSNR decreases with the increase in the number of unconnected pixels. As for multiple connected pixels in the to-be-highpass-filtered frame, the effect is reversed.

Typically, there are only about 3-5% of the pixels that will be "unconnec-ted" in the MCTF process [33]. Therefore, almost the same amount of noise after the synthesis is transformed from the uncorrected quantization noise to the reconstructed frames F_1 and F_2 via the MCTF. Even though the original frames are used as the reference frames in the MCTF at the encoder side and only partial information of the frames are available at the decoder side, there is no serious drift at the decoder side.

As a comparison, we also study the error propagation pattern in the conventional closed-loop based ME/MC, which is described by

$$H_1(m,n) = F_2(m,n) - F_1(m+1,n),$$
$$L_1(m+1,n) = F_1(m+1,n).$$

Assume that uncorrected noise terms σ_L^2 and σ_H^2 are applied to L_1 and H_1, respectively. The resulting noise terms are

$$\sigma_{F_1(m+1,n)}^2 = \sigma_L^2, \tag{9.43}$$
$$\sigma_{F_2(m,n)}^2 = \sigma_L^2 + \sigma_H^2. \tag{9.44}$$

Clearly, there is a serious drift problem associated with the conventional ME/MC scheme when the reference frames is the original ones.

The comparison between the MCTF and the conventional ME/MC scheme (CMC) is illustrated in Table. 9.2. Clearly, the MCTF is more suitable for the scalable video coding.

Table 9.2. The comparison between the MCTF and the conventional ME/MC

items	MCTF	CMC
reference frames	original frames	reconstructed frames
prediction gain	high	low
steps	prediction and up-date steps	prediction step
error propagation pattern	no serious drift pro-blem	drift problem
delay	depend on the GOF size	≤ 1

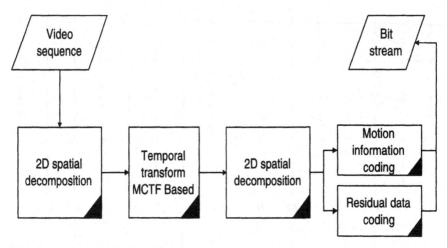

Fig. 9.14. A typical 3D subband video coding system

9.3.3 Scalable Video Coding Based on the MCTF

In the wavelet based 3D subband video coding, four types of redundancy are removed. They are temporal redundancy, spatial redundancy, perceptual redundancy and statistical redundancy. A typical example for 3D subband video coding is illustrated in Fig. 9.14. At the encoder side, the input video is first divided into GOPs. To provide the desired temporal scalability, the frames in each GOP are temporally decomposed by the MCTF. Suppose that the GOP size is 16, then the four stages of the analysis operation are recursively performed on the low temporal frequency subband to generate decomposition. As shown in Fig. 9.15 [116], after temporal decomposition, each GOP contains 16 frames: one t-LLLL frame, one t-LLLH frame, two t-LLH frames, four t-LH frames, and eight t-H frames. Five levels of temporal scalability is presented. At each round of MCTF, a low subband and a high subband are usually generated for each motion compensation pair using a rate distortion optimization with the utilization of a Lagrangian multiplier (λ_{opt}) in the existing schemes. Each λ_{opt} corresponds to a bit rate range and a tradeoff between the motion information and the residual information. The tradeoff is most crucial for an SVC scheme. A large λ_{opt} corresponds to a low bit rate while a small one corresponds to a high bit rate. Normally, the optimal point of an SVC scheme is the point where the first residual image is generated for each motion compensation pair. There only exists one optimal point for an SVC scheme.

After preforming all necessary MCTF, a spatial transform will be performed on each subband to make the energy unevenly distributed and to gather energy by the relative importance to the HVS. The spatial redundancy is then removed. Generally, there are two types of spatial transforms, discrete

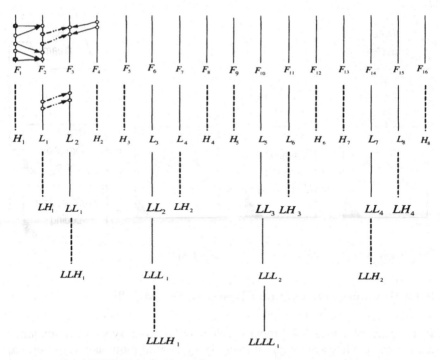

Fig. 9.15. An MACTF Based on HAAR

cosine transform (DCT) and discrete wavelet transform (DWT). The $N \times N$ DCT is

$$c(k,n) = \begin{cases} \frac{1}{\sqrt{N}}, & k = 0, 0 \le n \le N-1 \\ \sqrt{\frac{2}{N}} \cos(\frac{\pi(2n+1)k}{2N}), & 1 \le k \le N-1, 0 \le n \le N-1 \end{cases} \tag{9.45}$$

The DWT is a very nice tool to provide the spatial scalability. An example is illustrated in Fig. 9.16 where four levels of spatial scalability is provided. After the spatial transform, the residual image will be converted from the spatial domain to the frequency domain. If the invertible DCT or DWT is used in the spatial transform, it is lossless. Otherwise, it is lossy [77, 3]. We usually adopt Daubechies's 9/7 [3] or 5/3 analysis/synthesis filters for spatial filtering. The analysis and synthesis filters are given in Tables 9.3-9.5, respectively.

To remove the perceptual redundancy, the residual image in the frequency domain are usually quantized by a quantization matrix. The quantization matrix is designed according to one important feature of the HVS, that is, the human eyes are more sensitive to low frequency than to high frequency. Thus, a small element is chosen for the residual information at low frequency while a larger one is selected for that at high frequency. The quantization

Table 9.3. Low pass filters taps for floating point Daubechies (9,7) filter-band

taps	analysis	synthesis
0	0.60294901823636	1.115087052457
±1	0.266864118442875	-0.59127176311425
±2	-0.07822326652899	-0.0575435262285
±3	-0.016864118442875	0.09127176311425
±4	0.02674875741081	

Table 9.4. High pass filters taps for floating point Daubechies (9,7) filter-band

taps	synthesis	taps	analysis
1	0.60294901823636	-1	1.115087052457
0,2	0.266864118442875	-2,0	-0.59127176311425
-1,3	-0.07822326652899	-3,1	-0.0575435262285
-2, 4	-0.016864118442875	-4,2	0.09127176311425
-3, 5	0.02674875741081		

Table 9.5. Analysis and synthesis filters taps for (5,3) filter-band

taps	low pass (analysis)	low pass (synthesis)	taps	high pass (analysis)	taps	low pass (synthesis)
0	3/4	1	-1	1	1	3/4
±1	1/4	1/2	-2, 0	-1/2	0,2	-1/4
±2	-1/8				-1,3	-1/8

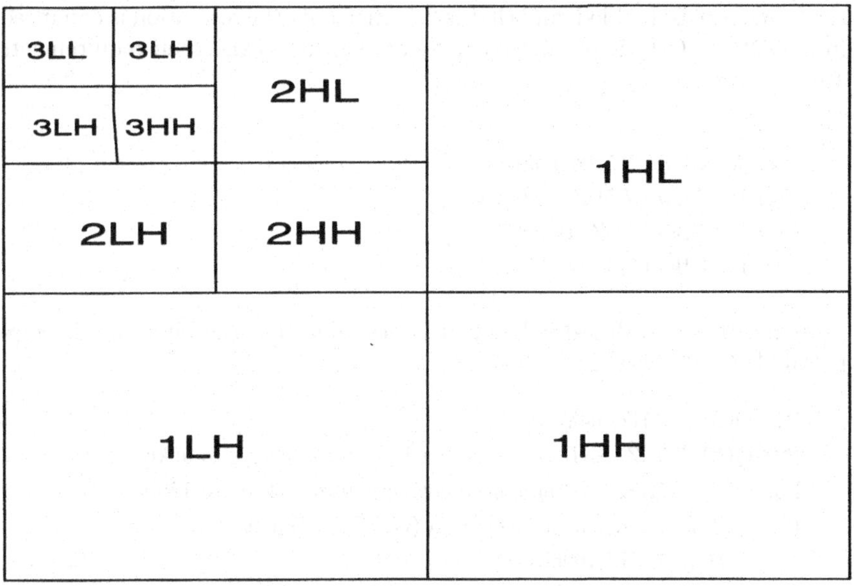

Fig. 9.16. An example of spatial scalability

process is usually lossy. The SNR scalability is achieved by properly choosing different quantization step at different bit rate.

After the above three steps, the residual image and the motion information are available for the final entropy coding. The entropy coding is used to remove the statistical redundancy among the residual image and the motion information. The essence of entropy coding is to use short symbols to represent values that occur frequently, and long symbols to represent values that occur less frequently. There are usually two types of entropy coding, variable length coding (VLC) and arithmetic coding. Normally, more statistical redundancy can be removed if the entropy coding is adaptive to the content of a video sequence. The entropy coding is lossless.

9.4 Multiple Adaptation Feature

Once the motion information and the residual information are available, it is very important to organize them such that several sub-bitstreams can be easily extracted. Meanwhile, the redundancy among them is minimized. An example is presented below.

Example 9.4.1. Assume that the increment factor of bit rate is 1.5 when only the frame rate is doubled, and that is 2.5 when only the resolution is enhanced from QCIF to CIF. Suppose that there are four users and their requirements are

> User A: QCIF, 7.5HZ, 64kb/s,
> User B: QCIF, 7.5HZ, 128kb/s,
> User C: CIF, 7.5HZ, 160kb/s,
> User D: CIF, 15HZ, 360kb/s.

To reduce the overall redundancy, it is desirable that the bitstream is composed of several parts as follows:

> P1: QCIF, 7.5HZ, 64kb/s,
> P2: QCIF,7.5HZ, SNR enhancement layer from 64kb/s to 96kb/s,
> P3: QCIF,7.5HZ, SNR enhancement layer from 96kb/s to 128kb/s,
> P4: spatial enhancement layer from (QCIF,7.5HZ,64kb/s)
>   to (CIF,7.5HZ,160kb/s),
> P5: temporal enhancement layer from P1, P2 and P4
>   to (CIF,15HZ,360kb/s).

The bitstreams sent to different users are given respectively as

User A: P1,

User B: P1+P2+P3,

User C: P1+P4,

User D: P1+P2+P4+P5.

□

The above example adopts the bottom-up approach. The bitstream is upg-
raded from the lower resolution and the lowest quality. The up-down approach
can also be used to achieve the same objective. For example, a sub-bitstream
of (CIF,15HZ,480kb/s) is first extracted from the whole bitstream, the sub-
bitstreams of (CIF,7.5HZ,320kb/s) and (CIF,15HZ,360kb/s) are extracted
from (CIF,15HZ,480kb/s). The sub-bitstreams of (CIF,7.5HZ,160kb/s) and
(QCIF, 7.5HZ, 128kb/s) are then extracted from (CIF,7.5HZ,320kb/s). Finally, the sub-bitstream of (QCIF, 7.5HZ, 64kb/s) is extracted from (QCIF,
7.5HZ, 128kb/s).

Clearly, the up-down approach is simpler than the bottom-up one but the
bandwidth may be wasted by the up-down approach.

Normally, it is desirable that the bitstreams have a multiple adaptation
features, that is, the extraction of a sub-bitstream in a pre-extracted sub-
bitstream should be comparable to the same extraction performed on the
entire bitstream wherever the extraction points are appropriately nested.

However, it is difficult to know in advance exactly whether one sub-
bitstream is embedded into another sub-bitstream. The concept of bits per
pixel (bpp) can be used to simplify it although it is not perfect. The bpp is
computed by

$$\text{bpp} = \frac{\text{bitrate (b/s)}}{\text{frame rate} * \text{frame size}} \tag{9.46}$$

For example, the sub-bitstream of (CIF, 15Hz, 128kbps) that has 0.08
bpp must be embedded into that of (CIF, 30Hz, 384kbps) with 0.13 bpp. It
is not necessarily embedded into that of (CIF, 30Hz, 192kbps) with 0.06 bpp.

Let $\Upsilon_1(i,j)$ and $\Upsilon_2(i,j)$ denote the lower and upper bounds of the SNR
range at resolution level i and temporal level j, respectively. To provide the
multiple adaptation feature, $\Upsilon_1(i,j)$ and $\Upsilon_2(i,j)$ need to satisfy

$$\Upsilon_1(1,j) \leq \Upsilon_1(2,j) \leq \cdots, \leq \Upsilon_1(J,j) \; ; \; \forall j \tag{9.47}$$
$$\Upsilon_1(i,1) \leq \Upsilon_1(i,2) \leq \cdots, \leq \Upsilon_1(i,L) \; ; \; \forall i \tag{9.48}$$
$$\Upsilon_2(1,j) \leq \Upsilon_2(2,j) \leq \cdots, \leq \Upsilon_2(J,j) \; ; \; \forall j \tag{9.49}$$
$$\Upsilon_2(i,1) \leq \Upsilon_2(i,2) \leq \cdots, \leq \Upsilon_2(i,L) \; ; \; \forall i \tag{9.50}$$

where J and L are the total number of resolution levels and SNR levels, respectively.

Meanwhile, arbitrary quality level within an SNR range should be reached for a video with any given resolution and frame-rate. If necessary, more enhancement in the SNR dimension is sent to augment with what can already be extracted at the newly adapted resolution and/or frame rate. However, the granularity has to be well controlled so as not to cause too much overhead in terms of pointers into the bitstream. This is shown in Example 9.4.1.

9.5 Switched Scalable Video Coding Schemes

9.5.1 A Customer Oriented Scalable Video Coding Scheme

Currently, SVC schemes have achieved excellent coding efficiency comparable to that of the state-of-the-art single-layer coding at the bit rate that the residual images are generated. Specifically, the residual images are generated at a low bit rate in the FGS by using a closed-loop ME/MC schemes, and the performance of the FGS is comparable to that of the single-layer coding at the low bit rate [197]. While the residual images are generated at a high bit rate in the MCTF based SVC and the performance is comparable to that of the single-layer coding at the high bit rate [142, 33]. However, the coding performance of the MCTF based SVC (or the FGS) is relatively poor at low (or high) bit-rates. There exists a PSNR gap about 1-3 dB compared with the state-of-the-art single layer coding scheme at very low (or high) bit rate.

Meanwhile, the coding efficiency of spatial scalable coding is significantly worse than that the state-of-the-art single layer coding. Specifically, the coding efficiency of high-resolution signal is very poor by the bottom-up approaches [166]. For the class of top-down approaches, the problem is contrary [200]. One of the major reasons is that the motion information and the residual information at different resolutions are generated independently. As a result, not all information that is coded in a base layer can be used for encoding the spatial enhancement layer even though the inter-layer prediction can be used to reduce the redundancy between successive spatial resolutions [166].

Furthermore, the existing schemes are not optimal from the commercial point of view. Normally, different users have different customer compositions and the SVC scheme should be optimized at the corresponding point determined by the customer composition instead of always at the high bit rate. The profit of the user can be maximized in this way. An example is given below.

Table 9.6. Customer composition of companies A and B

Company	QCIF, 7.5f/s, 64kb/s	CIF, 15f/s, 512kb/s	4CIF, 60f/s, 2Mb/s
A	100000	2M	10000
B	10000	100000	1.5M

Example 9.5.1. Assume that there are companies A and B with the customer composition given in Table 9.6.

Obviously, the optimal point should be chosen at (CIF, 15f/s, 512kb/s) for company A and (4CIF, 60f/s, 2Mb/s) for company B, respectively.   □

Based on the observation that a good commercial model is very important for the success of an SVC scheme, a grid ME/MC scheme is proposed in this section. The grid is determined by the granularity of each scalability, the optimal point and the suboptimal points chosen by the user according to the customer composition. The key idea of our scheme is to first guarantee the coding efficiency of the most important range and then to maximize the coding efficiency of other ranges according to their relative importance. Our scheme is described in detail as follows: the ME/MC starts with λ_{opt} corresponding to the optimal point, which generates a residual image and the corresponding motion information. They are called the basic motion information and the basic residual image. To guarantee the performance of the optimal point, the motion information and the residual image for the lower bit rate at the same resolution and the corresponding bit rate at the lower resolution are truncated from the basic motion information and the residual image. The truncations adopt a simple truncation scheme based on the RDO with a Lagrangian multiplier corresponding to the bit rate range. To maximize the performance of the suboptimal points, a novel scalable video coding scheme, i.e. cross layer ME/MC scheme, is proposed. In each ME/MC, the corresponding residual image and motion information are generated for the remaining coding process with a high bit rate range. Meanwhile, a switching law of motion information is defined in this section. The full motion information scalability is thus provided by our scheme. Obviously, our scheme is different from the existing ones in the sense that only one residual image is generated for each motion compensation pair at each resolution by the existing ones. However, two residual images may be generated for each motion compensation pair by our scheme. Our scheme can thus be regarded as a switched SVC scheme. The major difference between our proposed scheme and the existing ones is illustrated in Figs. 9.17 and 9.18.

Compared to the existing SVC schemes, ours has the following four major advantages:

1. Design a customer-oriented scalable video coding scheme as Fig. 9.19. In our scheme, the granularity of each scalability, the optimal point and the suboptimal points are chosen by the user according to his/her customer

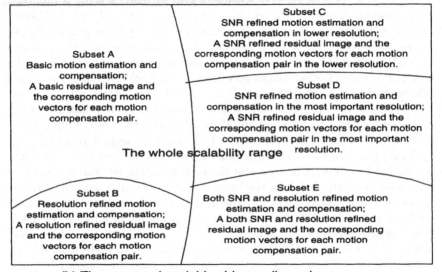

Fig. 9.17. The comparison between our scheme and the existing ones

profile. Our scheme first guarantees the coding efficiency of the most important range and then try to maximize the coding efficiency of other ranges according to their relative importance.

2. Improve the performance of the SNR scalability by proposing a cross SNR layer ME/MC scheme with the introduction of a new SNR ME/MC refinement criterion. Both the motion information and the residual information are refined by our cross SNR layer ME/MC scheme. We try to obtain a good trade-off between the motion information and the residual information at all bit rates.

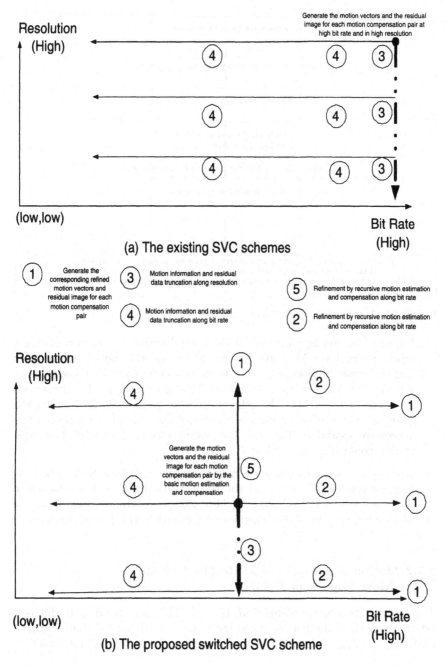

Fig. 9.18. The comparison between our scheme and the existing ones

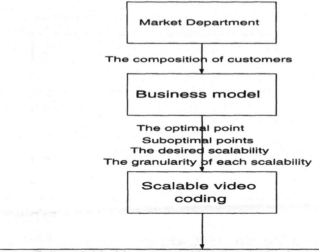

Fig. 9.19. A customer oriented SVC scheme

3. Improve the coding efficiency of the spatial scalability by introducing a cross spatial layer ME/MC scheme. Note that the motion information and the residual information at a higher resolution include those that are already coded in the lower resolution. This fact will be used to design our cross spatial layer ME/MC scheme. A new spatial ME/MC refinement criterion will also be introduced to reduce the redundancy between two successive spatial resolutions. The coding efficiency of spatial scalable coding could then be improved.

4. Present a new template, i.e. a hybrid template, for the SVC. Both the original reference frames and the reconstructed frames are involved in our cross layer ME/MC scheme. Our scheme is thus a hybrid template of an open-loop ME/MC scheme and a closed-loop ME/MC scheme.

9.5.2 Motion Information and Residual Image at the Most Important Point

Three parameters are predefined for the ME/MC corresponding to different bit rate ranges at the lth temporal level and the jth spatial level. Suppose that they are $\lambda_{opt}(l, j)$, $\lambda_{low}(l, j)$ and $\lambda_{high}(l, j)$, respectively, and satisfy

$$\lambda_{low}(l,j) \geq \lambda_{opt}(l,j) \geq \lambda_{high}(l,j), \tag{9.51}$$

$$\lambda_{opt}(l,1) \geq \lambda_{opt}(l,2) \geq \cdots \geq \lambda_{opt}(l,J), \tag{9.52}$$

$$\lambda_{low}(l,1) \geq \lambda_{low}(l,2) \geq \cdots \geq \lambda_{low}(l,J), \tag{9.53}$$

$$\lambda_{high}(l,1) \geq \lambda_{high}(l,2) \geq \cdots \geq \lambda_{high}(l,J). \tag{9.54}$$

$\lambda_{opt}(l, j)$ corresponds to the most important range at the lth temporal level and the jth spatial level, which can be determined by the customer composition. An example is illustrated in Fig. 9.19.

Assume that the optimal point is at temporal level l_0 and spatial level $\varphi(l_0)$. Define an optimal Lagrangian multiplier $\hat{\lambda}_{opt}(l, j)$ as

$$\hat{\lambda}_{opt}(l, j) = \begin{cases} \lambda_{high}(l, j); & \text{if } l < l_0 \\ \lambda_{opt}(l_0, \varphi(l_0)); & \text{if } l \geq l_0, j \leq \varphi(l_0) \\ \lambda_{opt}(l, j); & \text{otherwise} \end{cases} \tag{9.55}$$

At the temporal level $l(l = 1, 2, 3, \cdots, L)$, a resolution $\varphi(l)$ is chosen as the most important resolution by the user according to the customer composition. The ME/MC starts from the resolution $\varphi(l)$ with the parameter $\hat{\lambda}_{opt}(l, \varphi(l))$ at the lth temporal level, and is called the basic ME/MC at the lth temporal level.

To guarantee coding efficiency of the most important range, the motion information for the bit rate lower than the most important range is truncated from those obtained by the basic ME/MC. Similarly, at the resolution lower than $\varphi(l)$, the motion information for the bit rate range corresponding to $\hat{\lambda}_{opt}(l, \varphi(l))$ and those lower than it are scaled down and truncated from those at the resolution $\varphi(l)$. To maximize the coding efficiency of other ranges according to their relative importance, a cross SNR layer ME/MC scheme and a cross spatial layer ME/MC scheme are proposed.

9.5.3 SNR Scalability of Motion Information and Residual Image

The ME/MC starts from the bit rate range that corresponds to the parameter $\hat{\lambda}_{opt}(l, \varphi(l))$. The SNR scalability of motion information and residual image is guaranteed by the items listed below.

9.5.3.1 Motion Information Truncation at Low Bit Rate

At each resolution, the RDO with the parameter $\lambda_{low}(l, j)$ is used to generate the motion information at the low bit rate. The motion information is truncated from that obtained by using the basic ME/MC at the lth temporal level.

Assume that $F_2(x, y)$ is the reference picture. (dx_0, dy_0) is obtained by using the basic ME/MC scheme at the lth temporal level. Define a set of MVs around (dx_0, dy_0) as

$$\rho(dx_0, dy_0, \delta_x, \delta_y) = \{(dx, dy) | |dx - dx_0| < \delta_x, |dy - dy_0| < \delta_y\}. \tag{9.56}$$

The MV (dx, dy) is in the set $\rho(dx_0, dy_0, \delta_{xtb}, \delta_{ytb})$ and is the truncated one from (dx_0, dy_0). The ME/MC processes at the optimal point and the suboptimal point are illustrated in Figs 9.20 and 9.21, respectively. From the figures, an SNR truncation criterion is defined as

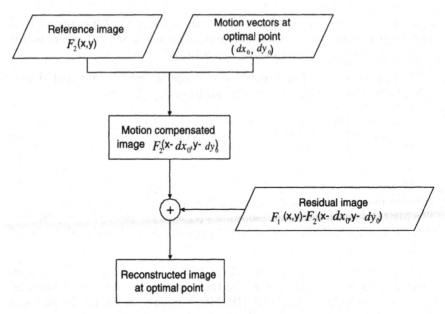

Fig. 9.20. The motion compensation at the optimal point

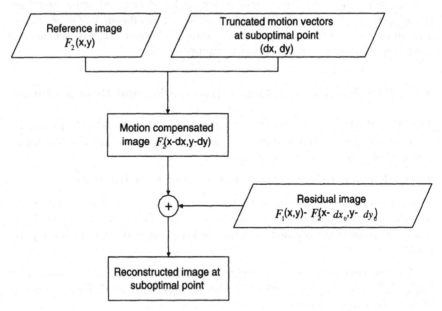

Fig. 9.21. The motion compensation at a suboptimal point

$$D_{bt}(dx, dy, dx_0, dy_0)$$
$$= \sum_{x,y} \max\{|F_2(x - dx, y - dy) - F_2(x - dx_0, y - dy_0)|$$
$$-P_r(F_2(x - dx_0, y - dy_0), F_1(x, y), \lambda_{low}(l, j)), 0\}, \tag{9.57}$$

where $P_r()$ is the perceptual redundancy determined by the background luminance masking, the bit rate, the temporal luminance changes and texture masking [207].

The MV (dx_0, dy_0) is truncated into (dx, dy) if the inequality

$$\lambda_{low}(l, j)R_{mv}(dx, dy) + D_{bt}(dx, dy, dx_0, dy_0) < \lambda_{low}(l, j)R_{mv}(dx_0, dy_0) \tag{9.58}$$

is satisfied.

The following result can be easily derived.

Proposition 9.5.1. *For two Lagrangian multipliers $\lambda_1 > \lambda_2$, if an MV (dx_0, dy_0) is truncated into (dx, dy) with λ_2, it will also be truncated into (dx, dy) with λ_1.* □

In other words, the set of MVs at a given bit rate is a subset of those at a relatively higher bit rate.

Note that a multi-layer motion coding scheme was proposed in [200] and the scheme is given as follows:

The base layer of MVF corresponds to a low bit rate and is generated by using a relatively larger λ, while the enhancement layers of MVF correspond to high bit rates and are generated by using a set of relatively small λ's [200]. The MVs at low bit rate are obtained by using (9.60) while the enhancement layers of MVF correspond to high bit rates and are generated using a set of relatively small λ's in [200]. An example is provided below.

Example 9.5.2. Assume that we require three levels of MVF scalability, they correspond to 128kb/s, 512kb/s and 2Mb/s, respectively. The λ's correspond to them are 300, 240, and 24, respectively. The MVFs that correspond to 128kb/s, 512kb/s and 2mb/s are generated by using three rate distortion optimization processes of the forms $R_{mv1} + 300 * D_1$, $R_{mv2} + 240 * D_2$ and $R_{mv3} + 24 * D_3$, respectively. It should be highlighted that D_1 and D_2 are only used to generate the MVFs at 128kb/s and 512kb/s. The residual image that is coded at 128kb/s and 512kb/s is D_3. Thus, there exists motion mismatch problem at low and medium bit rates. □

It can be known from (9.58) that

$$\lambda_{low}(l, j)R_{mv}(dx, dy) + D_{bt}(dx, dy, dx_0, dy_0) + D_p(dx_0, dy_0)$$
$$< \lambda_{low}(l, j)R_{mv}(dx_0, dy_0) + D_p(dx_0, dy_0). \tag{9.59}$$

That is,

$$\lambda_{low}(l,j)R_{mv}(dx,dy) + D_p(dx,dy)$$
$$< \lambda_{low}(l,j)R_{mv}(dx_0,dy_0) + D_p(dx_0,dy_0). \tag{9.60}$$

Clearly, if an MV is truncated in our scheme, it will also be pruned in the existing ones, while the reverse is not true. This implies that in the existing schemes, the motion information is not enough in the sense of the RDO at low bit rate. As a result, the motion mismatch may be serious and the coding efficiency is still very low at low bit rates. Our motion information truncation scheme is thus better than the multi-layer motion coding scheme.

Since we have two groups of motion information, we need to determine the switching point by

$$\max_{\lambda}\{R(\lambda)\}, \tag{9.61}$$

such that

$$\lambda R_{mv}(dx,dy) + D_{bt}(dx,dy,dx_0,dy_0) > \lambda R_{mv}(dx_0,dy_0), \tag{9.62}$$

where $R(\lambda)$ is the bit rate range corresponding to the parameter λ.

It should be highlighted that only the motion information to be coded, is switched.

9.5.3.2 Cross SNR Layer ME/MC Scheme at High Bit Rate

To maximize the coding efficiency at a high bit rate, both the coded residual information and the motion information at the optimal point should be used by the encoder to generate the motion information and the residual images at the high bit rate. Meanwhile, to speed up the ME/MC at high bit rate, block partition information and the corresponding motion information are shared by the ME/MC at different bit rate. There are many choices on the shared motion information. The final block partition mode can be fixed. In this case, the final block partition mode and the corresponding motion information are shared, and the number of MVs is the same at different bit rate. It is also possible to fix the initial full MV tree obtained in the ME/MC process. The initial full MV tree and the corresponding motion information are shared. The tree pruning processes are performed by using $\hat{\lambda}_{opt}(l,j)$ and $\lambda_{high}(l,j)$ from the same initial full MV tree. The number of MVs may not be the same at different bit rate. It is also possible to fix the final block partition mode at one resolution while fixing the initial full MV tree at another resolution.

Assume that $F_1(x,y)$ and $F_2(x,y)$ are the predicted frame and the reference frame, respectively, and the motion information MV_{opt} and the residual image $(F_1(x,y) - F_2(x - dx_0, y - dy_0))$ $((dx_0,dy_0) \in MV_{opt})$ are generated by using the basic ME/MC scheme at the l_0th temporal level. All motion information and an optimal part of the residual image $Q(F_1(x,y) - F_2(x - dx_0, y - dy_0))$ are coded and sent to the decoder at the optimal point. $Q()$ is the quantization operation, and the value of $Q()$ is determined by the customer composition.

When the bit rate is higher than the optimal point, the remaining energy $D_{br}(dx_0, dy_0, dx_0, dy_0)$ (defined in equation (9.63)) should be coded and sent to the decoder. There are two possible methods to code $D_{br}(dx_0, dy_0, dx_0, dy_0)$ at higher bit rate. One way is to code it directly, which is the same as the FGS scheme. Actually, this is the INTRA coding mode of the remaining energy. The other is to further perform ME/MC, to generate a new residual image and additional MVs, and to code the new residual image and the additional MVs. This is an INTER coding mode of the remaining energy.

Note that the motion information is too much at low bit rate when it is generated at high bit rate, just as in the existing SVC schemes. Conversely, MV_{opt} may not be enough at higher bit rate. Thus, the latter one should be better than the former one. Meanwhile, the maximum coding gain is nearly achieved by minimizing the signal energy of the temporal H-subband, since the energy of the temporal L-subband is relatively constant. From these, we propose our cross SNR layer ME/MC scheme as below.

Both the motion information and the coded residual information $Q(F_1(x, y) - F_2(x - dx_0, y - dy_0))$ are involved in the refinement. This is illustrated in Fig. 9.22.

Let (dx_0, dy_0) denote an MV obtained by the basic ME/MC scheme at temporal level l. (dx, dy) is in the set $\rho(dx_0, dy_0, \delta_{xrb}, \delta_{yrb})$ and is a candidate for the refinement. Define an SNR refinement criterion as

$$D_{br}(dx, dy, dx_0, dy_0) = \sum_{x,y} |F_1(x,y) - F_2(x - dx, y - dy)$$
$$-IQ(Q(F_1(x,y) - F_2(x - dx_0, y - dy_0)))|. \qquad (9.63)$$

With the new criterion defined in equation (9.63), a cross SNR-layer ME/MC scheme is available to improve the performance of the SNR scalability [112].

Suppose that $R_{mv}(dx - dx_0, dy - dy_0)$ is the number of bits to code the MV $(dx - dx_0, dy - dy_0)$. A further ME/MC with an MV as (dx, dy) will be performed at the lth temporal level and the $\varphi(l)$th spatial level if the inequality

$$D_{br}(dx, dy, dx_0, dy_0) + \lambda_{high}(l, \varphi(l))R_{mv}(dx - dx_0, dy - dy_0)$$
$$\leq D_{br}(dx_0, dy_0, dx_0, dy_0) \qquad (9.64)$$

is satisfied.

Let (dx_1, dy_1) and (dx_2, dy_2) be two refined MV candidates of (dx_0, dy_0). (dx_1, dy_1) will be chosen if

$$D_{br}(dx_1, dy_1, dx_0, dy_0) + \lambda_{high}(l, \varphi(l))R_{mv}(dx_1 - dx_0, dy_1 - dy_0)$$
$$\leq D_{br}(dx_2, dy_2, dx_0, dy_0) + \lambda_{high}(l, \varphi(l))R_{mv}(dx_2 - dx_0, dy_2 - dy_0). \qquad (9.65)$$

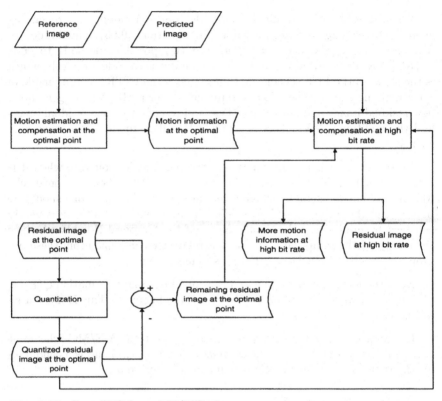

Fig. 9.22. Cross SNR layer ME/MC scheme

Note that the inequality in the existing ME/MC scheme is given by

$$D_p(dx, dy) + \lambda_{high}(l, \varphi(l))R_{mv}(dx - dx_0, dy - dy_0) \leq D_p(dx_0, dy_0). \qquad (9.66)$$

The only difference between $D_{br}(dx, dy, dx_0, dy_0)$ in equation (9.63) and $D_p(dx, dy)$ in equation (9.3) is that an additional item $IQ(Q(F_1(x, y) - F_2(x - dx_0, y - dy_0)))$ is involved in $D_{br}(dx, dy, dx_0, dy_0)$. The inequality (9.64) can be regarded as a generalized version of the inequality (9.66).

Similar to the SVC scheme in [165], there are many coding codes for the remaining energy. The ME/MC is the same as that in Subsection 9.2.3 and the mode decision process for our cross SNR layer ME/MC scheme is the same as that in [165] except that the performance indices (9.3) and (9.7) are replaced by those defined in (9.64) and (9.65).

There are certain constraints on the choices of δ_{xrb} and δ_{yrb} when the cross SNR layer ME/MC scheme is adopted in an SVC that is based on the MCTF. There is no constraint when the cross SNR layer ME/MC scheme is adopted in an SVC that is on basis of the unconstrained MCTF [185].

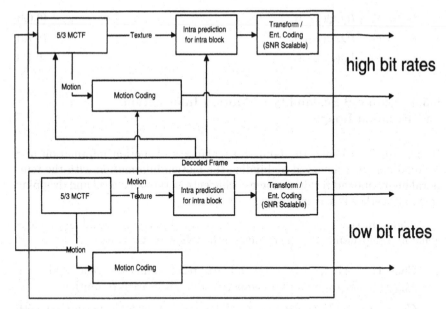

Fig. 9.23. An example of our cross SNR layer ME/MC scheme

Example 9.5.3. Our proposed cross SNR layer ME/MC scheme can also be used to improve the coding efficiency of the existing FGS schemes at high bit rate [197]. This is demonstrated in Fig. 9.23. □

Similarly, we need to determine at which point we should switch when the cross SNR layer ME/MC scheme is used. The switching point satisfies

$$\max_{\lambda}\{R(\lambda)\}, \tag{9.67}$$

such that

$$D_{br}(dx, dy, dx_0, dy_0) + \lambda R_{mv}(dx - dx_0, dy - dy_0)$$
$$< D_{br}(dx_0, dy_0, dx_0, dy_0). \tag{9.68}$$

It should be highlighted that both the motion information and the residual image to be coded are switched when the cross SNR layer ME/MC scheme is used.

The coded information should be removed to improve the coding efficiency. Assume that the residual images at three resolutions after the ME/MC with $\lambda_{opt}(l, j)$ and $\lambda_{high}(l, j)$ are E_{11}, E_{21}, E_{31}, $\hat{F}_1$, $\hat{F}_2$ and $\hat{F}_3$, respectively. This is shown in Fig. 9.27. The redundancies between $IQ(Q(E_{11}))$ and $\hat{F}_1$, $IQ(Q(E_{21}))$ and $\hat{F}_2$, and $IQ(Q(E_{31}))$ and $\hat{F}_3$ should be removed. The residual images E_{12}, E_{22} and E_{32} for the coding at high bit rate are thus computed by

$$E_{12} = \hat{F}_1 - IQ(Q(E_{11})), \tag{9.69}$$
$$E_{22} = \hat{F}_2 - IQ(Q(E_{21})), \tag{9.70}$$
$$E_{32} = \hat{F}_3 - IQ(Q(E_{31})). \tag{9.71}$$

9.5.4 Temporal Scalability of Motion Information and Residual Image

In our MCTF scheme, the temporal scalability of motion information and residual images is achieved by binding the motion information with the temporal high-subband residual image coefficients at the same level and dropping them altogether if it is necessary.

To guarantee the performance of the optimal point, the residual image is generated depending on l according to the following two cases:

Case 1: $l \in \{1, 2, \cdots, l_0 - 1\}$. Only one residual image is generated using $\lambda_{high}(l, j)$ for each motion compensation pair at each resolution.

Case 2: $l \in \{l_0, l_0 + 1, \cdots, L\}$. The ME/MC will be performed with two parameters, $\hat{\lambda}_{opt}(l, j)$ and $\lambda_{high}(l, j)$. Thus, there are two groups of motion information, and two residual images generated for each motion compensation pair at the jth resolution. They will be used to generate the motion information and the residual images that will be coded at each range.

Thus, two temporal low and high subbands are generated corresponding to the two constants at temporal level $l(l = l_0 + 1, l_0 + 2, \cdots, L)$. The remaining MCTFs at those levels are performed with the corresponding low subbands.

The ME/MC starts with $\lambda_{opt}(l, j)$, $F_1(x, y)$ and $F_2(x, y)$. We should take full advantage of the MVs and the block partition modes obtained by using $\lambda_{opt}(l, j)$ when we perform ME/MC with $\lambda_{high}(l, j)$, $F_1(x, y)$ and $F_2(x, y)$. The coded MVs of a block in the previous layer can be used as one candidate of predicted MV and the block partition modes obtained from the previous layer can be regarded as the initial status of partitioning for the co-located block.

9.5.5 Spatial Scalability of Motion Information and Residual Image

In most conventional wavelet-based scalable video coding, spatial scalability is realized with the wavelet transform. Lower resolution sequence is generated by reducing the DWT level of higher resolution motion predicted error. In this case, the motion information of lower resolution sequence is generated by dividing that of higher resolution by 2, while macroblock size should be halved for both horizontal and vertical directions. Obviously, lower resolution sequence is always affected by large amount of motion information. This is the main reason why conventional schemes have low visual quality or even cannot produce the bit-stream at lower resolution sequence. In addition, the mismatch between the halved motion information and the motion predicted error of reduced size which is a low subband of the higher resolution error leads to severe performance degradation. As a result, even at very high bit-rates, the visual quality of the lower resolution image cannot reach the level of the original reference image that is generated by wavelet-transform.

To overcome this, the MCTF starts from and optimized at the most important resolution $\varphi(l)$. Cross spatial layer ME/MC can be performed for the resolution higher than the most important one. At resolution j lower than $\varphi(l)$, only one ME/MC with the parameter $\lambda_{high}(l,j)$ is performed. At the resolution $\varphi(l)$, another ME/MC with the parameter $\lambda_{high}(l,\varphi(l))$ is further performed to refine the residual image. At resolution j higher than $\varphi(l)$, one ME/MC with $\lambda_{opt}(l,j)$ and another one with $\lambda_{high}(l,j)$ are performed to generate the optimal motion information and residual images for the corresponding bit rate ranges. In each ME/MC, the corresponding residual image and motion information are generated for the remaining coding process at the corresponding bit rate range. For the sake of simplicity, we assume that there are a total of three resolutions and $\varphi(l) = 2$.

Suppose that the reference frames are $\tilde{F}_2(x/2, y/2)$ and $F_2(x,y)$ at the middle resolution and the highest resolution, respectively. $\tilde{F}_2(x/2,y/2)$ can be $S_D(F_2(x,y))$ with S_D as a down-sampling operation. $\tilde{F}_2(x/2,y/2)$ and $F_2(x,y)$ can also be independent. The motion information and the residual images are generated by the basic ME/MC scheme at the lth temporal level. All motion information and an optimal part of residual information $Q(S_D(F_1(x,y)) - \tilde{F}_2(x/2 - dx_0, y/2 - dy_0))$ are sent to the decoder at the middle resolution.

The spatial scalability of motion information and residual images is obtained through the processes described below.

9.5.5.1 Motion Information Truncation at the Lowest Resolution

At the lowest resolution, the MVs are scaled down and truncated from those obtained by the basic ME/MC scheme at the lth temporal level. All MVs are first scaled down by a factor of 2 and the resolution down by half. The final block partition mode at the middle resolution serves as the initial full MV tree at the lowest resolution. The tree pruning process is performed by using the $\hat{\lambda}_{opt}(l,1)$ to generate the final block partition mode.

Assume that (dx, dy) is in the set $\rho(dx_0, dy_0, \delta_{xts}, \delta_{yts})$ and is the truncated MV of MV (dx_0, dy_0) where (dx_0, dy_0) is obtained by using the basic ME/MC scheme at the lth temporal level as in Figure 9.27. A spatial truncation criterion is

$$D_{st}(dx, dy, dx_0, dy_0)$$
$$= \sum_{x,y} \max\{|S_D(\tilde{F}_2(x/2 - dx, y/2 - dy)) - S_D(\tilde{F}_2(x/2 - dx_0, y/2 - dy_0))|$$
$$-P_r(S_D(\tilde{F}_2(x/2 - dx_0, y/2 - dy_0)), S_D(S_D(F_1(x,y))), \lambda_{low}(l,1)), 0\}. \quad (9.72)$$

The MV (dx_0, dy_0) is truncated into (dx, dy) if the inequality

$$\hat{\lambda}_{opt}(l, 1)R_{mv}(dx, dy) + D_{st}(dx, dy, dx_0, dy_0) < \hat{\lambda}_{opt}(l, 1)R_{mv}(dx_0, dy_0) \quad (9.73)$$

is satisfied.

9.5.5.2 Cross Spatial Layer ME/MC Scheme at the Highest Resolution

The motion information and the coded residual information are shared between the middle resolution and the highest resolution to speed up the ME/MC and to maximize the coding efficiency at the highest resolution. To achieve these, both the coded residual information and all motion information at the middle resolution should be shared with and used by the encoder to generate the residual information and the motion information at the highest resolution.

The motion information and the residual images at the highest resolution are generated by using ME/MC between $F_2(x, y)$ and $F_1(x, y)$. Since the motion information and the residual information at the highest resolution include those at the middle resolution, the motion information MV_{opt} and the coded residual information $Q(S_D(F_1(x, y)) - \tilde{F}_2(x/2 - dx_0, y/2 - dy_0))$ should be involved in the ME/MC at the highest resolution. This scheme is called cross spatial layer ME/MC scheme. The redundancy between two successive spatial resolution can then be minimized by the scheme.

Let (dx_0, dy_0) denote an MV obtained by the basic ME/MC scheme at the lth temporal level. (dx, dy) is in the set $\rho(2dx_0, 2dy_0, \delta_{xrs}, \delta_{yrs})$ and is a candidate for the refinement. Define a spatial refinement criterion as [111]

$$D_{sr}(dx, dy, dx_0, dy_0) = \sum_{x,y} |F_1(x, y) - F_2(x - dx, y - dy)$$
$$-S_U(IQ(Q(S_D(F_1(x, y)) - \tilde{F}_2(x/2 - dx_0, y/2 - dy_0))))|, \quad (9.74)$$

where S_U is an up-sampling operation.

With the new criterion defined in equation (9.74), a cross spatial layer ME/MC scheme can be introduced to improve the performance of the spatial scalability. Meanwhile, it can be known from equations (9.64) and (9.74) that both the original reference frames and the reconstructed frames are involved in our cross layer ME/MC scheme. Thus, our scheme is a hybrid template of both the open-loop ME/MC scheme and the closed loop ME/MC scheme.

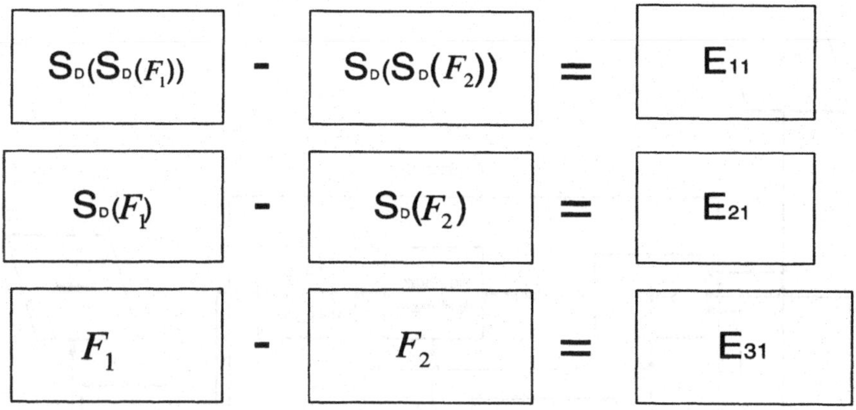

$S_D(S_D(F_1))$	$-$	$S_D(S_D(F_2))$	$=$	E_{11}
$S_D(F_1)$	$-$	$S_D(F_2)$	$=$	E_{21}
F_1	$-$	F_2	$=$	E_{31}

Fig. 9.24. Encoder side of spatial scalability

$C(E_{11})$	$+$	$S_D(S_D(F_2))$	$=$	$S_D(S_D(F_1))$
$C(E_{21})-$ $S_U(S_D(E_{21}))$	$+$	$S_D(F_2)-$ $S_U(S_D(S_D(F_2)))$	$=$	$S_D(F_1)-$ $S_U(S_D(S_D(F_1)))$
$C(E_{31})-$ $S_U(S_D(E_{31}))$	$+$	$F_2-S_U(S_D(F_2))$	$=$	F_1- $S_U(S_D(F_1))$

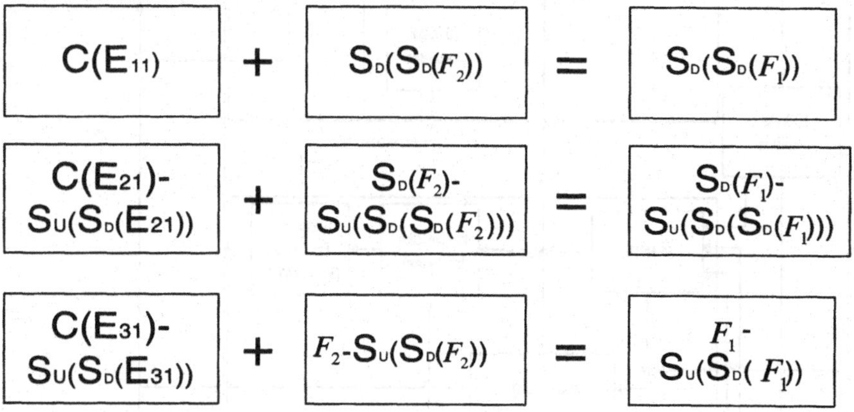

Fig. 9.25. Decoder side of spatial scalability

A further ME/MC with an MV as (dx, dy) will be performed at the highest resolution if the inequality

$$D_{sr}(dx, dy, dx_0, dy_0) + \lambda_{opt}(l, 3)R_{mv}(dx - 2dx_0, dy - 2dy_0)$$
$$\leq D_{sr}(2dx_0, 2dy_0, dx_0, dy_0) \tag{9.75}$$

is satisfied [112].

Assume that (dx_1, dy_1) and (dx_2, dy_2) are two refined MV candidates of (dx_0, dy_0). MV (dx_1, dy_1) will be chosen if

$$D_{sr}(dx_1, dy_1, dx_0, dy_0) + \hat{\lambda}_{opt}(l, 3)R_{mv}(dx_1 - 2dx_0, dy_1 - 2dy_0)$$
$$\leq D_{sr}(dx_2, dy_2, dx_0, dy_0) + \hat{\lambda}_{opt}(l, 3)R_{mv}(dx_2 - 2dx_0, dy_2 - 2dy_0). \tag{9.76}$$

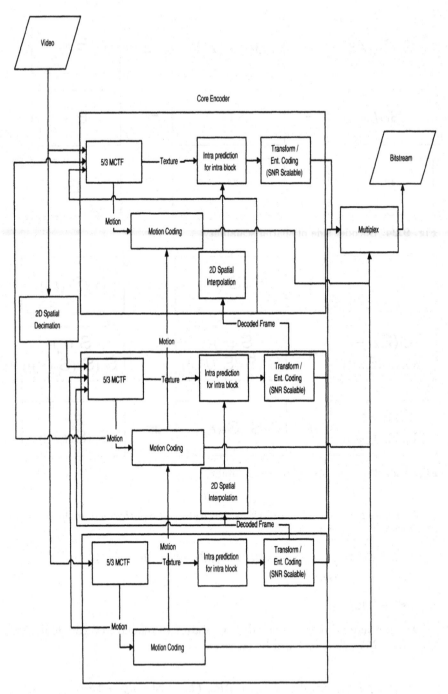

Fig. 9.26. An example of our cross spatial layer ME/MC scheme

Similarly, there are certain constraints on the choices of δ_{xrs} and δ_{yrs} when the cross spatial layer ME/MC scheme is adopted in an SVC that is based on the MCTF (or the MCTF and the UMCTF). There is no constraint when the cross spatial layer ME/MC scheme is adopted in an SVC that is on basis of the unconstrained MCTF [185].

The coded residual images $IQ(Q(E_{11}))$, $IQ(Q(E_{21}))$ and $IQ(Q(E_{31}))$ in Figure 9.27 is

$$IQ(Q(E_{11})) = IQ(Q(S_D(E_{21}))),$$
$$IQ(Q(E_{21})) = IQ(Q(E_{21} - S_U(IQ(Q(E_{11}))))) + S_U(IQ(Q(E_{11}))), \qquad (9.77)$$
$$IQ(Q(E_{31})) = IQ(Q(E_{31} - S_U(IQ(Q(E_{21}))))) + S_U(IQ(Q(E_{21}))). \qquad (9.78)$$

The corresponding residual image and motion information are generated for the remaining coding process at the corresponding resolution. The process is illustrated in Figs. 9.24 and 9.25 in the case that $\tilde{F}_2 = S_D(F_2)$.

Similarly, there are many coding codes at a higher resolution. The ME/MC is the same as that in Subsection 9.2.3 and the mode decision process for our cross spatial layer ME/MC scheme is the same as that in [165, 166] except that the performance indices (9.3) and (9.7) are replaced by those defined in (9.75) and (9.76).

Example 9.5.4. Consider Example 9.1.2 again. Suppose that $\varphi(l) = 1$. The encoder structure of this example is illustrated in Fig. 9.26. Our encoder outperforms that in [165, 166] in the sense that both the motion information and the coded residual information at a lower resolution are used by our encoder to generate the motion information and the residual information at a higher resolution. All coded information at a lower resolution can then be used for encoding the spatial enhancement layers. The coding efficiency can thus be improved at the higher resolution. □

The process in Example 9.5.4 can be shown in the following example.

Example 9.5.5. The most important resolution is the middle one and $\lambda_{opt}(l,j)$ corresponds to low bit rate. The whole process on the tradeoff between the motion information and the residual information is illustrated in Fig. 9.27 at temporal level 1.

Clearly, the motion mismatch problem associated with the existing scalable motion information coding scheme is totally eliminated when $\lambda_{opt}(l,j)$ corresponds to low bit rate, the most important resolution is the low resolution at each temporal level as illustrated in Fig. 9.28. □

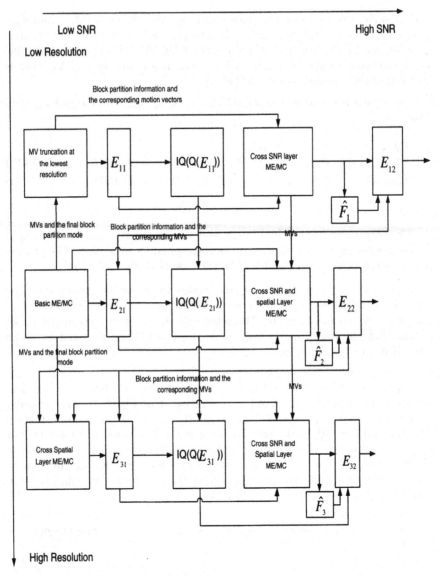

Fig. 9.27. Our proposed grid ME/MC scheme

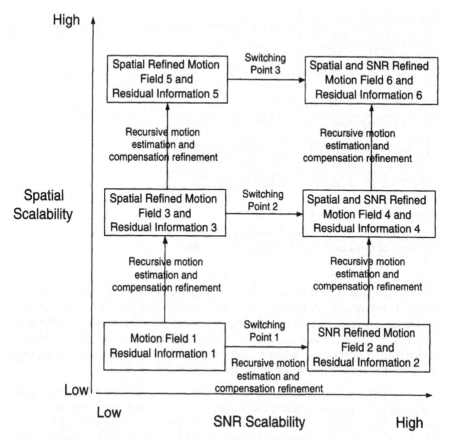

Fig. 9.28. Our cross layer ME/MC scheme

Example 9.5.6. **Customer Oriented Trade-off Scheme**

Four standard video sequences, Bus, Mobile, Foreman and Football with CIF size (352×288) [51], are used in our experiments to compare our proposed scheme with the scheme in [200]. The frame rates of these sequences are set as 30f/s. Each temporal level has two layers of MVFs and they are generated with the values of λ_{low} and λ_{opt} given in Table 9.7 and the testing points are listed in Table 9.8.

Our experiments are based on the software provided by the Microsoft [136]. Ends and Starts are two parameters that should be chosen in advance.

The following method is used to extract the MVFs for the scheme in [200]. The motion information is enhanced smoothly together with the residual information. The Ends of Bus, Mobile and Football are the same and they are (1,1,1,1), (1,1,1,0), (1,1,0,0), and (0,0,0,0), while those of Foreman are (1,1,1,1), (1,1,1,0), (1,1,0,0), (1,0,0,0), (0,0,0,0), and (0,0,0,0).

Table 9.7. The values of λ_{low} and λ_{opt}

temporal level	1	2	3	4
λ_{low}, QCIF	135	135	135	135
λ_{opt}, CIF	16	32	64	64

Table 9.8. The testing points

Video Sequence	Testing points(kb/s)
Foreman (CIF, 30f/s)	48, 64, 96, 128, 160, 2048
Football (CIF, 30f/s)	128, 160, 192, 2048
Bus (CIF, 30f/s)	128, 160, 192, 2048
Mobile (CIF, 30f/s)	128, 160, 192, 2048

Table 9.9. The customer profile of user C

bit rate	low	medium	high
number of customers	2M	200,000	10,000

Assume that the customers of user C is given in Table 9.9. All $\lambda_{opt}(l)(l = 1, 2, 3, 4)$ are always chosen as 135 in our scheme and the Ends are (1,1,1,1) and Starts=Ends=1 at all fourth temporal levels. We also test the case where $\lambda_{opt}(l)(l = 1, 2, 3, 4)$ are chosen as 16, 32, 64, and 64, respectively. This is called the conventional scheme and Starts=Ends=0 at the first temporal level, and Starts=Ends=1 at other three temporal levels.

We shall test our trade-off scheme between the motion information and the residual information when the values of both ν_1 and ν_2 in equation (9.20) are 1's. The experimental results on the peak signal-to-noise ratio (PSNR) of the Y component are shown in Tables 9.10-9.13 where $PSNR$ is computed by

$$PSNR = 10 \log_{10} \frac{255^2}{\sigma_e^2}, \tag{9.79}$$

σ_e^2 is the mean squared error (MSE) between the original video sequence and the reconstructed video sequence. It can be shown that our scheme can be used to improve the PSNR by up to 1.68dB at the optimal point.   □

Example 9.5.7. **The Effect of the Values of** (ν_1, ν_2)

We shall test the values of (ν_1, ν_2) in equation (9.20). It is the conventional 5/3 MCTF when both their values are 1's. The results are listed in tables 9.14-9.17 and Figs. 9.29 and 9.30. Obviously, the PSNR variation is reduced by up to 29% where the average PSNR is slightly reduced when they are chosen as (32/33,5/4). It is also shown that both the average PSNR and the variation are increased with an increase in ν_1 within a range.   □

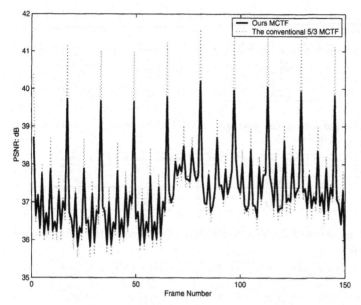

Fig. 9.29. Comparison of the PSNR variation between (1,1) and (32/33,5/4) for the sequence Bus

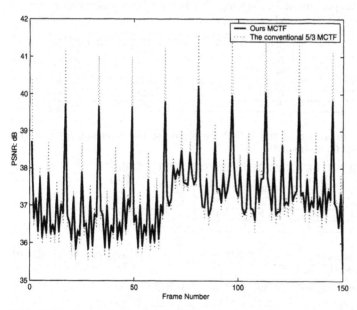

Fig. 9.30. Comparison of the PSNR variation between (1,1) and (32/33,5/4) for the sequence Foreman

Table 9.10. Comparison of the average PSNR for the sequence Mobile with different schemes, GOFsize=64

Bit Rate(kb/s)	Layered motion scheme in [200]	Conventional scheme	Ours
128	23.36	23.28	23.65
160	24.47	24.3	24.53
192	25.28	25.23	25.23
2048	37.33	37.24	36.62

Table 9.11. Comparison of the average PSNR for the sequence Football with different schemes, GOFsize=64

Bit Rate(kb/s)	Layered motion scheme in [200]	Conventional scheme	Ours
128	24.29	16.58	25.97
160	25.13	21.54	26.56
192	26.09	24.43	27.05
2048	37.82	38.4	37.26

Table 9.12. Comparison of the average PSNR for the sequence Bus with different schemes, GOFsize=64

Bit Rate(kb/s)	Layered motion scheme in [200]	Conventional scheme	Ours
128	24.42	24.56	25.24
160	25.56	25.72	25.98
192	26.46	26.58	26.52
2048	38.19	38.36	37.41

Table 9.13. Comparison of the average PSNR for the sequence Foreman with different schemes, GOFsize=64

Bit Rate(kb/s)	Layered motion scheme in [200]	Conventional scheme	Ours
48	27.28	19.12	27.81
64	28.38	24.11	28.88
96	29.84	29.67	30.31
128	30.9	31.26	31.25
160	31.51	32.32	31.98
2048	41.7	42.27	41.06

Table 9.14. Comparison of the PSNR variation for the sequence Mobile with different schemes, GOFsize=64

Bit rate (kb/s)	PSNR $(1, 1)$	STDEV $(1, 1)$	PSNR $(\frac{32}{33}, \frac{5}{4})$	STDEV $(\frac{32}{33}, \frac{5}{4})$	PSNR $(\frac{16}{15}, \frac{5}{4})$	STDEV $(\frac{16}{15}, \frac{5}{4})$
128	23.65	0.32	23.62	0.32	23.7	0.34
160	24.53	0.41	24.51	0.36	24.61	0.41
192	25.23	0.42	25.26	0.41	25.29	0.48
2048	36.62	1.72	36.57	1.47	36.72	1.84

Table 9.15. Comparison of the PSNR variation for the sequence Football with different schemes, GOFsize=64

Bit rate (kb/s)	PSNR $(1, 1)$	STDEV $(1, 1)$	PSNR $(\frac{32}{33}, \frac{5}{4})$	STDEV $(\frac{32}{33}, \frac{5}{4})$	PSNR $(\frac{16}{15}, \frac{5}{4})$	STDEV $(\frac{16}{15}, \frac{5}{4})$
128	25.97	3.09	25.94	2.99	26.01	3.17
160	26.56	3.08	26.51	2.97	26.58	3.18
192	27.05	3.09	27.02	3	27.1	3.15
2048	37.26	3.04	37.09	2.87	37.31	3.14

Table 9.16. Comparison of the PSNR variation for the sequence Bus with different schemes, GOFsize=64

Bit rate (kb/s)	PSNR (1, 1)	STDEV (1, 1)	PSNR $(\frac{32}{33}, \frac{5}{4})$	STDEV $(\frac{32}{33}, \frac{5}{4})$	PSNR $(\frac{16}{15}, \frac{5}{4})$	STDEV $(\frac{16}{15}, \frac{5}{4})$
128	25.24	0.81	25.2	0.7	25.41	0.73
160	25.98	0.7	25.94	0.62	26.01	0.95
192	26.52	0.74	26.57	0.66	26.58	1.01
2048	37.41	1.31	37.3	0.94	37.53	1.4

Table 9.17. Comparison of the PSNR variation for the sequence Foreman with different schemes, GOFsize=64

Bit rate (kb/s)	PSNR (1, 1)	STDEV (1, 1)	PSNR $(\frac{32}{33}, \frac{5}{4})$	STDEV $(\frac{32}{33}, \frac{5}{4})$	PSNR $(\frac{16}{15}, \frac{5}{4})$	STDEV $(\frac{16}{15}, \frac{5}{4})$
48	27.81	1.07	27.8	1.04	27.89	1.14
64	28.88	1.16	28.86	1.16	28.96	1.14
96	30.31	1.14	30.24	1.09	30.37	1.18
128	31.25	1.12	31.19	1.09	31.33	1.2
160	31.98	1.12	31.85	1.07	32.08	1.16
2048	41.06	1.33	40.92	1.02	41.16	1.4

9.6 Low Delay Scalable Video Coding Schemes

There are many cases where a low delay SVC scheme is desired. A typical example is given as follows:

Example 9.6.1. **Surveillance systems**

Images from a set of surveillance cameras can be transmitted to a number of different users. Some of these users may be at a fixed location (e.g. a designated CCTV room), some may be mobile, but within wireless LAN range (e.g. a security guard with a handheld device), and still others may be at a remote location (e.g. a home owner checking the security of their residence with a 2.5G/3G mobile phone). Different users have different requirements as described below:

1. Central surveillance room connected to LAN prefers clear picture and smooth motion video stream because available bandwidth is high enough.

2. Remote surveillance room connected to narrowband network prefers clear picture than motion smoothness because its terminal display size is big enough to show fine resolution of picture but available bandwidth is low.

3. Security guard with mobile terminal connected to wireless network prefers smooth motion and error-robust video stream because its terminal display size is small and network has error prone characteristics.

The SVC can be used to satisfy the above requirements simultaneously. Meanwhile, clients may also want to control camera parameter(pan/tilt) to follow the criminal immediately after the object is detected. If the delay is too long, the criminal will be missed out in the recording in the database. The same requirement comes from real-time video conference systems. The end-to-end delay should be less than 500 ms in this case. So scalable video coding should support a low delay mode.   □

Assume that the GOF size is n. The delay associated with each ME/MC scheme is listed in Table 9.18.

Table 9.18. The delay of each ME/MC scheme

MC scheme	delay
the conventional one	≤ 1
the Haar MCTF	$2^n - 1$
the 5/3 MCTF	$2^{n+1} - 2$
the 3-band MCTF	$\frac{3^n - 1}{2}$

To generate a bitstream that contains a low delay sub-bitstream, the MCTF and the conventional ME/MC schemes should be seamlessly integrated by combining their advantages listed in Table 9.2. An example is used to illustrate the process.

Example 9.6.2. Suppose that the input frame rate is 30HZ and the frame size is 4CIF. The user is more interested in the QCIF size video sequence with 7.5HZ for low delay. The encoder can generate a desired picture for the coding of low delay as follows: first it performs 2 rounds of temporal wavelet transform on 4 successive original pictures to generate a video sequence with 4CIF size and 7.5 HZ. Then it performs 2 rounds of two dimensional spatial wavelet transform to generate a desired video sequence. The video sequence is then coded via an H.264 encoder.

It then generates the enhancement layer bitstream, and performs all necessary MCTFs on the original pictures with 4CIF size and the predefined parameters in each GOF to generate low and high subbands. The reconstructed pictures in the H.264 coding process are used to perform further rounds of MCTF to generate the corresponding low and high subbands. The redundancies among them will be removed via the method similar to the above. Code the remaining information by using the existing methods for the MCTF.   □

Therefore, the most important part is to design a proper ME/MC scheme among frames in a GOF when a low delay bitstream is necessary. In the following, we shall provide an example to show how to design the ME/MC scheme. Even though we are dealing with motion compensated lifting scheme, for the sake of simplicity, we shall write in the sequel of the corresponding prediction equations without taking into account of the motion aspect.

Example 9.6.3. Assume that the size of a GOF is chosen as 32 for the low delay bitstream, the original frames are denoted by F_i $(i = 1, 2, \cdots, 32)$. The desired temporal scalability for the low delay is 30Hz, 15Hz and 7.5Hz, respectively. The tolerated delay is 2/15s. Suppose that the generated frames after the ME/MC are denoted by H_i $(1 \leq i \leq 32)$. The MC scheme is

$$\begin{bmatrix} H_1 \\ H_2 \\ H_3 \\ \vdots \\ H_{32} \end{bmatrix} = \begin{bmatrix} A_{11} & 0 & 0 & \cdots & 0 \\ A_{21} & A_{22} & 0 & \cdots & 0 \\ A_{31} & A_{32} & A_{33} & \cdots & 0 \\ \vdots & \vdots & \vdots & \ddots & \vdots \\ A_{81} & A_{82} & A_{83} & \cdots & A_{88} \end{bmatrix} \begin{bmatrix} F_1 \\ F_2 \\ F_3 \\ \vdots \\ F_{32} \end{bmatrix}, \tag{9.80}$$

where $A_{ij} \in R^{4 \times 4}$.

Note that there is no drift problem associated with the MCTF and the delay can be well controlled by the unconstrained MCTF [57] or the conventional closed-loop based ME/MC scheme. It is thus desirable to use the MCTF as much as possible and to integrate the MCTF with the unconstrained MCTF seamlessly when a low delay bitstream is provided. The following method can be used to achieve this objective.

1. All $A_{ii}(i = 1, 2, \cdots, 8)$ are determined by the MCTF.

2. All other A_{ij} are determined by the MCTF and the unconstrained MCTF (or the conventional closed-loop based ME/MC scheme) together.

Let

$$A = \begin{bmatrix} A_{11} & 0 & 0 & \cdots & 0 \\ A_{21} & A_{22} & 0 & \cdots & 0 \\ A_{31} & A_{32} & A_{33} & \cdots & 0 \\ \vdots & \vdots & \vdots & \ddots & \vdots \\ A_{81} & A_{82} & A_{83} & \cdots & A_{88} \end{bmatrix}. \tag{9.81}$$

A necessary and sufficient condition for perfect reconstruction is that all matrices A_{ii} $(i = 1, 2, \cdots, 8)$ are non-singular [187]. The video sequence is reconstructed by

$$F = A^{-1}H. \tag{9.82}$$

An interesting case is that all A_{ii} $(i = 2, 3, \cdots, 8)$ are the same. An example is illustrated in Fig. 9.31. Here, the solid line represents a low subband or an original frame, a dash line represents a high subband or a residual image.

The 5/3 MCTF is chosen in our design. All A_{ij} are equal to 0, except the matrices given below:

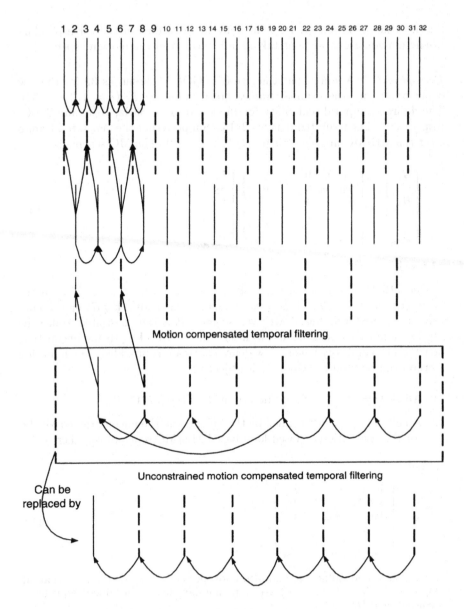

Motion compensated temporal filtering

Unconstrained motion compensated temporal filtering

Can be
replaced by

Closed loop based motion compensation

Fig. 9.31. A low delay example

$$A_{11} = \begin{bmatrix} 1 & -1 & 0 & 0 \\ 0 & -\frac{1}{2} & 1 & -\frac{1}{2} \\ \frac{1}{4} & \frac{7}{8} & -\frac{1}{4} & -\frac{7}{8} \\ \frac{1}{8} & \frac{3}{16} & \frac{3}{8} & \frac{5}{16} \end{bmatrix},$$

$$A_{ii} = \begin{bmatrix} 1 & -\frac{1}{2} & 0 & 0 \\ 0 & -\frac{1}{2} & 1 & -\frac{1}{2} \\ \frac{1}{4} & 1 & -\frac{1}{4} & -\frac{7}{8} \\ \frac{1}{8} & \frac{1}{4} & \frac{3}{8} & \frac{5}{16} \end{bmatrix} \; ; \; i = 2, 3, 4, 5, 6, 7, 8,$$

$$A_{i(i-1)} = \begin{bmatrix} 0 & 0 & 0 & -\frac{1}{2} \\ 0 & 0 & 0 & 0 \\ 0 & 0 & 0 & -\frac{1}{8} \\ -\frac{1}{8} & -\frac{3}{16} & -\frac{3}{8} & -\frac{3}{8} \end{bmatrix} \; ; \; i = 2, 3, 4, 6, 7, 8,$$

$$A_{54} = \begin{bmatrix} 0 & 0 & 0 & -\frac{1}{2} \\ 0 & 0 & 0 & 0 \\ 0 & 0 & 0 & -\frac{1}{8} \\ 0 & 0 & 0 & -\frac{1}{16} \end{bmatrix},$$

$$A_{51} = \begin{bmatrix} 0 & 0 & 0 & 0 \\ 0 & 0 & 0 & 0 \\ 0 & 0 & 0 & 0 \\ -\frac{1}{8} & -\frac{1}{4} & -\frac{3}{8} & -\frac{5}{16} \end{bmatrix}.$$

Note that there is a drift problem associated with the unconstrained MCTF, it is necessary to control the length of the drift by defining the matrices. In the above example, the length is set as 4. □

10. Future Research Directions and Potential Applications

Many interesting applications of switched and impulsive systems have been presented in the previous chapters. We shall conclude this book by identifying several future directions and potential applications in regards to switched and impulsive systems. We hope that they are useful as a guide to interested readers when exploring the system theory and applications of switched and impulsive systems.

Application 1: Global stabilization of nonlinear system with abrupt change.

Switched and impulsive control could be an efficient method to provide the global stabilization of nonlinear systems with abrupt change. In these cases, each sub-controller is designed to work well in a local area and the whole switched controller is designed such that it can work well in the whole space [104]. With a well designed switching law, the system will change from one area to another area, and the corresponding local controller is also switched into action. This process repeats until the state approaches its equilibrium.

Application 2: Control of light railway transfer (LRT) system.

An LRT system is a classic switched system. A train has a stationary trajectory but it has to stop at every station. Thus, comparing with an automated highway system (AHS), the discrete part of an LRT system is simpler while the continuous part is more complex. Thus, it is necessary to study an LRT system even though there are many results on AHS systems. When an LRT system is studied, a simple way is to find a leading train, whose control law can be totally determined by its own state. Then, the control law of the train following the leading train can be determined according to the requirements, and in turn the control law for the immediate next train. The process is repeated until the control law of every train is obtained.

Application 3: Congestion control of local area network (LAN).

It is possible to set up a switched model for a LAN when the congestion control problem of a LAN is considered. There are already some existing results for both the cases of one congestion point and multiple congestion

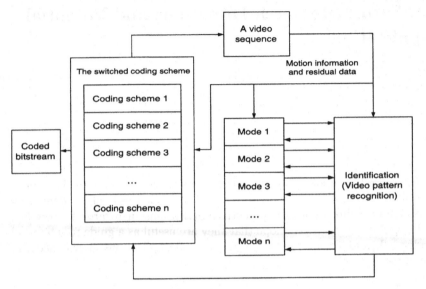

Fig. 10.1. A switched video coding scheme

points. However, the situations for a LAN are not fixed. Indeed, in practice, these situations are always changing. So, to exercise more effective congestion control, a group of sub-controllers should be developed. At any time, the sub-controller corresponding to the right situation should be used.

Application 4: Next generation of video coding system.

Switched control can be used to achieve better performance for nonlinear systems with abrupt changes than a continuous control. We note that there are also abrupt changes in a video sequence. For example, one part of a video sequence can be with low motion and texture, while another part is with high motion but low texture, yet the other part is with high motion and texture. It is thus necessary to design a switched coding scheme for the conventional non-scalable video coding system to obtain a higher coding efficiency. One example is illustrated in Fig. 10.1.

Application 5: Switched deblocking scheme for scalable video coding.

The block artifacts occur when the block-based pseudo motion vector instead of the true motion at every pixel is used in the encoding and decoding. There are two types of block artifacts, one is directly caused by the reference pixels, the other is caused by different error propagation pattern on the blocks in high and low subbands.

In most systems, each block motion vector is used in encoding and decoding for uniform block-wise displacement, and for predicting the value of the entire current block of pixels by the value of a displaced block from

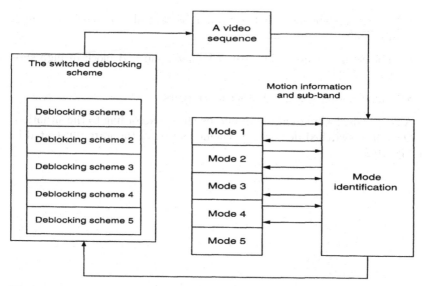

Fig. 10.2. A switched deblocking scheme

the previous frame. The block based motion estimation and compensation uses an implicit assumption that each block of pixels moves with uniform translational motion. However, the actual motion field in real world video sequences usually varies smoothly within the same moving object but are discontinuous at boundaries of different moving objects. The boundaries of a moving object are most likely to be inconsistent to the block boundaries. Therefore, the block based method is well known to produce block artifacts. Overlapped block motion compensation is an efficient method to reduce/eliminate this type of block artifacts [147].

A block in a high subband is called an I block if neither a prediction step nor an update step is performed on it, a P block if only a prediction step is performed on it, a B block if both a prediction step and an update step are performed on it. A block in a low subband is called an I block if no update step is performed on it, a P block if an update step is performed on it. Since error is propagated among different blocks following different patterns, there are block artifacts among different types of blocks. To improve the visual quality, it is desired to design a deblocking scheme to reduce/eliminate the second type of block artifacts.

Five different deblocking schemes can be designed according to the following five cases:

- Deblocking scheme 1, between an I block and a P block in a low subband;

- Deblocking scheme 2, between an I block and a P block in a high sub-band;

- Deblocking scheme 3, between an I block and a B block in a high sub-band;

- Deblocking scheme 4, between a P block and a B block in high sub-band.

- Deblocking scheme 5, all other block pairs.

A proper deblocking scheme should be chosen according to the situation. Overall, a switched deblocking scheme can be of the form illustrated in Fig. 10.2.

References

1. Alur R., Courcoubetis C., Dill D. (1993) Model Checking in Dense Real Time. Information and Computation 104: 2–34.
2. Alur R., Dill D. (1994) A Theory of Timed Automata. Theoretical Computer Science 126: 183–235.
3. Antonimi M., Barland M., Mathieu P., Daubechies I. (1992) Image Coding Using Wavelet Transform. IEEE Trans. on Image Processing 1: 205–200.
4. Antsaklis P. J., Lemmon M.D. Editors (1998) Special Issue on Hybrid Systems. Journal on Discrete Event Dynamic Systems: Theory and Applications 8: 99–239.
5. Antsaklis P. J., Nerode A. Editors (1998) Special Issue on Hybrid Control Systems. IEEE Transactions on Automatic Control 43: 457–587.
6. Antsaklis P. J. Editor (2000) Special Issue on Hybrid Systems: Theory and Applications. Proceedings of the IEEE 88: 879–1123.
7. Arulambalam A., Chen X. Q., Ansari N. (1997) An Intelligent Explicit Rate Control Algorithm for ABR Service in ATM Networks. In IEEE ICC'97. 200–204.
8. Au James (2003) Context Adaptive Variable Length Decoding System and Method. United State Patent. 6646578.
9. Back A., Guckenheimer J., Myers M. (1993) A Dynamic Simulation Facility for Hybrid Systems. In: Grossman R., Nerode A., Ravn A., Rischel H. (Eds) Hybrid Systems. 29–53.
10. Bainov D. D., Simeonov P. S. (1989) Systems with Impulse Effect: Stability, Theory and Applications, Halsted Press, New York.
11. Bencze W. J., Franklin G. F. (1995) A Separation Principle for Hybrid Control System Design. IEEE Control Systems Magazines 15: 80–84.
12. Benmohamed L., Meerkov S. M. Feedback Control of Congestion in Packet Switching Networks: the Case of a Single Congested Node. IEEE/ACM Transactions on Networking 1:693–707.
13. Benvensite A., Guernic P. (1990) Hybrid Dynamic Systems and the Signal Language. IEEE Transactions on Automatic Control 35: 535–546.
14. Bharghavan V., Lee K. W., Lu S. W., Ha S. W., Li J. R., Dwyer D. (1998) TIMELY Adaptive Resourece Management Architecture. IEEE Personal Communications5:20–31.
15. Blondel V., Tsitsiklis J. N. (1997) NP-Hardness of Some Linear Control Design Problems. SIAM J. Control Optim. 35: 2118–2127.
16. Boyd S., Ghoui L. EI, Feron E., Balakrishaan V. (1994) Linear Matrix Inequalities in System and Control Theory, SIAM, Boston.
17. Brave Y., Heymann M. (1990) Stabilization of Discrete–Event Process. International Journal of Control. 51: 1101–1117.
18. Branicky M. S. (1994) Stability of Switched and Hybrid Systems. In: Proc. of the 33nd CDC. 269–279.

19. Branicky, M. S. (1996) On a Class of General Hybrid Dynamic Systems. In: the 13th IFAC World Congress, USA, June 30–July 5, 287–292.
20. Branicky M. S., Borkar V. S., Mitter S. K. (1998) A Unified Framework for Hybrid Control: Model and Optimal Control Theory. IEEE Transactions on Automatic Control 43: 31–45.
21. Branicky M. S. (1998) Multiple Lyapunov Functions and Other Analysis Tools for Switched and Hybrid Systems. IEEE Transactions on Automatic Control 43: 475–482.
22. Brockett R. W. (1983) Asymptotic Stability and Feedback Stabilization. In: R. W. Brockett et al. (eds) Differential Geometric Control Theory. 181–191.
23. Brockett R. W. (1993) Hybrid Models for Motion Control Systems. In: Trentelman H. L. and Willems J. C. (Eds) Essays on Control Perspectives in the Theory and its Applications. 29–53.
24. Bruce Schneier (1994) Applied Cryptography, Protocols, Algorithm and Source Code in C, New York: J. Wiley.
25. Carlson M., Weiss W., Blake S., Wang Z., Black D. and Davis E. (1998) An Architecture for Differentiated Services, RFC 2475, December.
26. Chao H. J. (2002) Next Generation Routers. Proceedings of the IEEE 90:1518–1558.
27. Chen C. (1984) Linear System Theory and Design. Rinehart and Winston, New York.
28. Chen C., Soh Y. C., Li Z. G. (2005) Relative Differentiated Service Scheme: Dual Time Varying Deficit Round Robin. Submitted to Computer Network.
29. Chen C., Li Z. G., Soh Y. C. (2005) Source Adaptation for Real Time Applications Over Relative Differentiated Quality of Service. Submitted to IEEE Infocomm 2005.
30. Chen C., Li Z. G., Soh Y. C. (2005)Adaptive Video Streaming System Over Network with Relative Differentiated Service. Submitted to IEEE Transactions on Circuits and Systems for Video Technology.
31. Chen J., Koc T. V. Liu K. J. Ray (2001)Design of Digital Video Coding Systems. Marcel Dekker.
32. Chen M. S., HWang Y. R., Kuo Y. J. (2003) Exponentially Stabilizing Division Controller for Dyadic Bilinear Systems. IEEE Transactions on Automatic Control 48: 106–110.
33. Choi S. J., Woods John W. (1999) Motion-Compensated 3-D Subband Coding of Video. IEEE Transactions on Image Processing 8: 155-167.
34. Chua L. O., Desor C. A., Kuh E. S. (1987) Linear and Nonlinear Circuits, McGraw-Hill, New York.
35. Chua L. O. (1992) The Genesis of Chua's Circuit. Archivfur Elektronik and Ubertragungstechnik 46: 250–257.
36. Chua L. O., Itoh M., Kocarev L., Eckert K. (1993) Chaos Synchronization in Chua's Circuit. J. Circuit Syst. Comput. 3: 93–108.
37. Chua L. O. (1993) Global Unfolding of Chua's Circuit. IEICE Transactions on Fundament 76: 704–734.
38. Chua L. O. (1994) Chua's Circuit-an Overview Ten Years Later. J. Circuit Syst. Comput. 4: 117–159.
39. Clark D., Fang W. (1998) Explicit Allocation of Best–Effort Packet Delivery Service. IEEE/ACM Transactions Networking 6: 362–373.
40. Clark J., Holton D. A. (1991) A First Look At Graph Theory, World Scientific, Singapore.
41. Cover T. M., Thomas J. A. Elements of Information Theory. John Wiley & Sons, New York, 1991.

42. Cuomo M., Oppenheim V., Strogatz H. (1993) Synchronization of Lorenz-based Chaotic circuits, with Applications to Communications. IEEE Transactions on Circuits and Systems I: Fundamental Theory and Applications 40: 626–633.

43. David R., Alla H. (1987) Continuous Petri Nets. In : the 8th Eurpoean Workshop on Theory and Application of Petri Nets. 275–294.

44. Dayawansa W. P., Martin C. F. (1999) A Converse Lyapunov Theorem for a Class of Dynamical Systems with Undergo Switching. IEEE Transactions on Automatic Control 44: 751–760.

45. Devasia S., Paden R., Carlo R. (1997) Exact–Output Tracking for Systems with Parameter Jumps. International Journal of Control 67: 117–131.

46. Dovrolis C., Ramanathan P. (1999) A Case for Relative Differentiated Service and the Proportional Differentiation Model, IEEE/ACM Transactions On Network 13: 26–34.

47. Dovrolis C., Ramanathan P. (2000) Proportional Differentiated Service, Part II: Loss Rate Differentiation and Packet Scheduling, in IEEE/IFIP Int. Workshop Quality of Service (IWQoS). 52–61.

48. E. N. Lorenz. (1963) Deterministic Nonperiodic Flow. Journal of Atmospheric Science 20: 130–141.

49. Ezzine J., Haddad A. H. (1989) Controllability and Observability of Hybrid Systems. International Journal of Control 49: 2045–2055.

50. Floyd S., Jacobson (1993)Random Early Detection Gateways for Congestion Avoidance. IEEE/ACM Transactions on Networking 1: 397–413.

51. ftp://ftp.tnt.uni-hannover.de/pub/testsequences/svc, 2004.

52. Fu M., Barmish B. R. (1986) Adaptive Stabilization of Linear Systems via Switching Control. IEEE Transactions on Automatic Control 31: 1097–1103.

53. Gollu A., Varaiya P. P. (1989) Hybrid Dynamic Systems. In: Proc. of the 28th CDC, 2708–2712.

54. Gonzales A., Hee G., Gyvex P., Sinencio E. (2000) Lorenz-based Chaotic Cryptosystem: a Monolithis Implementation. IEEE Transactions on Circuits and Systems Part I: Fundamental Theory and Applications 8: 1243–1247.

55. Gunter M., Braun T. (2001) A Fast and Trend-Sensitive Function for the Estimation of Near-Future Data Network Traffic Characteristics. In: ASTC/ATS 2001. 22-26.

56. Halle K., Wu C. W., Itoh M., Chua L. O. (1992) Spread Spectrum Communication Through Modulation of Chaos. IEICE Transactions on Communicaton. 3: 469–477.

57. Han W. J. (2004) Response of Call-for-Proposal for Scalable Video Coding. ISO/IEC JTC1/SC 29 WG 11 MPEG2004/M10569/S07, Munchen, Germany.

58. Hang H. M., Tsai S. S., Chianh T. H. (2003) Motion information scalability for MC-EZBC: Response to Call for Evidence on Scalable Video coding. ISO/IEC JTC1/Sc29/WG11, MPEG2003/m9756, Tronheim.

59. Haskell B. G., Puri A., Netravali A. N. (1997) Digital Video: An Introduction to MPEG-2. Chapam Hall.

60. Hilhorst R. A., Amerongen J. V., öhnberg P. L., Tulleken H. J. A. F. (1994) A Supervisor for Control of Model-Switch Processes. Automatica 30: 1319–1331.

61. Hocherman-Frommer J., Kulkarni S. R., Ramadge P. J. (1998) Controller Switching Based on Output Prediction Errors. IEEE Transactions on Automatic Control 43: 596–607.

62. Holloway L. E., Krogh B. (1992) Properties of Behavioral Models for a Class of Hybrid Dynamic Systems. In : Proc. of the 31th CDC. 3752–3757.

63. Hou L., Michel A. N., Ye H. (1996) Stability Analysis of Switched Systems. In: Proc. of the 35th CDC. 1208–1212.

64. Hou L., Michel A. N., Ye H. (1997) Some Qualitative Properties of Sampled–Data Control Systems. In :Proc. of the 36th CDC. 911–916.
65. Hu B., Xu X., Michel A. N., Antsaklis P. J. (1999) Stability Analysis for a Class of Nonlinear Switched Systems. In: Proc. of the 38th CDC. 4374–4379.
66. Hu B., Xu X., Antsaklis P. J., Michel A. N. (1999) Robust Stabilizing Control Laws for a Class of Second–Order Switched Systems. System and Control Letters 38: 197–207.
67. Huang H. C., Wang C. N. Chiang T. H. (2002) A Robust Fine Granularity Scalability Using Trellis-Based Predictive Leak. IEEE Transaction on Circuits and Systems for Video Technology 12: 372-384.
68. Hunt B., Ott E., Yorke A. (1997) Differentiable Generalized Synchronism of Chaos. Physical View Letter 55: 4029–4033.
69. Hunt E. R., Johnson G. (1993) Keeping Chaos at Bay. IEEE Spectrum 32: 32–36.
70. Hurley P., Boudec J., Thiran P., Kara M. (2001) ABE: Providing a Low Delay Service Within Best Effort. IEEE Networks 15: 60–69.
71. Isidori A. (1996) Nonlinear Control Systems, Springer, London.
72. ISO/IEC. (1999) MPEG-4 Video Verification Model Version 13.0. ISO/IEC JTC 1/SC29/WG11 N2687.
73. ISO/MPEG. (1992) Coding of Moving Picture and Associated Audio for Digital Storage Media at up to 1.5Mbps. ISO/IEC 11172-2.
74. ISO/MPEG. (2000) Generic Coding of Moving Picture and Associated Audio Information. ISO/IEC 13818-2.
75. Jain R., Kalyanaraman S., Viswanathan R. (1994) Rate Based Control: Mistakes to Avoid. ATM Forum/94-0882 8:1–6.
76. Jain R. (1996) Congestion Control and Traffic Management in ATM Networks: Recent Advances and A Survey. Computer Networks amd ISDN Systems 28:462–476.
77. Jayant N. S., Noll P. (1984) Digital Coding of Waveforms. Prentice-Hall.
78. Ji Y., Li Z. G., Wen C. Y. (2005) Impulsive Synchronization of Chaotic Systems via Linear Matrix Inequalities. Submitted to International Journal of Bifurcaion and Chaos.
79. Johansson H., Rantzer A. (1998) Computation of Piecewise Quadratic Lyapunov Functions for Hybrid Systems. IEEE Transactions on Automatic Control 43: 555–559.
80. Kalman M., Steinback E., Girod B. (2004) Adaptive Media Playout for Low-Delay Video Streaming Over Error-Prone Channles. IEEE Transactions on Circuits and Systems for Video Technology 14: 841-851.
81. Khalil H. K. (1992) Nonlinear systems, Macmilan, New York.
82. Kohn W., James J., Nerode A., Harbison K., Agrawala A. (1995) A Hybrid Systems Approach to Computer-Aided Control Engineering. IEEE Control Systems Magazines 15: 14–25.
83. Kolarov A., Ramamurthy G. (1997) A Control Theoretic Approach to the Design of Closed Loop Rate Based Flow Control for High Speed ATM Networks. In Proceedings of IEEE INFOCOM'97. 293–301.
84. Kulkarni S. R., Ramadge P. J. (1996) Model and Controller Selection Policies Based on Output Prediction Errors. IEEE Transactions on Automatic Control 41: 1594–1604.
85. Lakshmikantham V., Bainov D. D., Simeonov P. S. (1989) Theory of Impulsive Differential Equations, World Scientific, London.
86. Lakshmikantham V., Liu X. Z. (1993) Stability Analysis in Terms of Two Measures, World Scientific, New York.

87. Lau F. C. M., Tse C. K. (2003) Chaos-Based Digital Communication Systems, Springer-Verlag.
88. LeBail J., Alla H., David R. (1991) Hybrid Petri Nets. In : Proc. of the 1st Euopean Control Conf. 1718–1724.
89. Lee D. C., Lough D. L., Midkiff S. F., Davis N. J., Benchoff P. E. (1998) The Next Generation of the Internet: Aspects of the Internet Protocol Version 6. IEEE Network 12:28–33.
90. Lemmon M. D., Antsaklis P. J. (1997) Timed Automata and Robust Control: Can We Now Control Complex Dynamic Systems? In: Proc. of the 36th CDC. 108–113.
91. Lemmon M. D., He K. X., Markovsky I. (1999) Supervisory Hybrid Systems. IEEE Control Systems Magazines 19: 42–55.
92. Liberzon D., Morse A. S. (1999) Basic Problems in Stability and Design of Switched Systems. IEEE Control Systems Magazines 19: 59–70.
93. Li W. P. (2001) Overview of Fine Granularity Scalability in MPEG-4 Video Standard. IEEE Transactions on Circuits and Systems for Video Technology 11: 301–317.
94. Li Z. G., Soh C. B., Xu X. H. (1997) Stability of Hybrid Dynamic Systems. In: Proc of the Second Asian Control Conference. 105-108.
95. Li Z. G., Soh C. B. , Xu X. H. (1997) Task–Oriented Design for Hybrid Dynamic Systems. International Journal of Systems Science. 28: 595–610.
96. Li Z. G., Soh C. B., Xu X. H. (1998) Robust Stability of a Class of Hybrid Dynamic Systems. International Journal of Robust and Nonlinear Control 8: 1059–1072.
97. Li Z. G., Wen C. Y., Soh Y. C. (1999) A Unified Approach for Stability Analysis of Impulsive Hybrid Systems. In: Proc. of the 38th CDC. 4398–4403.
98. Li Z. G., Wen C. Y., Soh Y. C. (1999) Stability of Perturbed Switch Nonlinear Systems. In: Proc. of the 1999 ACC. 2969–2974.
99. Li Z. G., Wen C. Y., Soh Y. C. (2000) Two State Transformations for Switched Systems. In: Proc. of the 2000 ASCC. 1596–1601.
100. Li Z. G., Wen C. Y., Soh Y. C. (2000) Robust Stability of Switched Systems with Arbitrarily Fast Switchings by Generalized Matrix Measure. In: Proc. of the 2000 ACC. 222–224.
101. Li Z. G., Soh C. B., Xu X. H. (2000) Lyapunov Stability for a Class of Hybrid Dynamic Systems. Automatica. 36: 297–302.
102. Li Z. G., Soh Y. C., Wen C. Y. (2000) Stability of Uncertain Quasi-Periodic Hybrid Dynamic Systems. International Journal of Control 73: 63–73.
103. Li Z. G., Soh Y. C., Wen C. Y. (2001) Robust Stability for a Class of Hybrid Nonlinear Systems. IEEE Transactions on Automatic Control 46: 897–903.
104. Li Z. G., Wen C. Y., Soh Y. C. (2001) Switched Controllers with Applications in Bilinear Systems. Automatica 37: 477–481.
105. Li Z. G., Wen C. Y., Soh Y. C. (2001) Analysis and Design of Impulsive Control Systems. IEEE Transactions on Automatic Control 46: 894–897.
106. Li Z. G., Wen C. Y., Soh Y. C., Xie W. X. (2001) The Stabilization and Synchronization of Chua's Oscillators via Impulsive Control. IEEE Transactions on on Circuits and Systems Part I: Fundamental Theory and Applications 40: 1351–1354.
107. Li Z. G., Yuan X. J., Wen C. Y., Soong B. H. (2001) A Switched Priority Scheduling Mechanism for ATM Switches with Multi-Class Output Buffers. Computer Networks 35: 203–221.
108. Li Z. G., Zhu C., Yang X. K., Feng G. N., Wu S., Pan F., Ling N. (2002) A Router Based Unequal Error Control for Video Over the Internet. In: IEEE 2002 International Conference on Image Processing. 713–716.

109. Li Z. G., Li K., Wen C. Y., Soh Y. C. (2003) A New Chaotic Secure Communication System. IEEE Transactions on Communications 51: 1306-1312.
110. Li Z. G., Zhu C., Ling N., Yang X. K., Feng G. N., Wu S., Pan F. (2003) A Unified Architecture for Real Time Video Coding Systems. IEEE Transactions on Circuits and Systems for Video Technology 13: 472-487.
111. Li Z. G., Rahardja S. (2004) Customer Oriented Scalable Video Coding. IEEE Transactions on Image Processing. Submitted.
112. Li Z. G., Yang X. K., Lim K. P., Lin X., Rahardaj S., Pan F. (2004) Customer Oriented Scalable Video Coding. ISO/IEC JTC1/SC29 WG11 MPEG2004/M11187, Palma de Mallorca.
113. Liberzon D. (1999) ISS and Integral–ISS Disturbance Attenuation with Bounded Controls. In: Proceedings of the 38th IEEE CDC. 2501–2506.
114. Lin A. M., Silvester J. A. (1991) Priority Queueing Strategies and Buffer Allocation Protocols for Traffic Control at an ATM Integrated Broadband Switching System. IEEE Journal on Selected Areas in Communications 9: 1524–1536.
115. Liu X. P., Wong W. S. (1997) Controllability of Linear Feedback Control with Communication Constraints. In: Proc. of the 36th CDC. 60–65.
116. Liu Y., Li Z. G., Soh Y. C (2004) Motion Compensation Temporal Filtering With Optimal Temporal Distance Between Each Motion Compensation Pair. In IEEE International Conference on Image Processing.
117. Lohmiller W., Slotine J. J. E. (1998) On Contraction Analysis of Non-linear Systems. Automatica 34: 683–696.
118. Lyapunov A. M. (1992) The General Problem of the Stability of Motion, Taylor and Francis, London.
119. Madan R. (1993) Special Issue on Chua's Circuit: A Paradigm for Chaos, Part I, Vol. 3 of J. Circuit, Systems, and Computers. World Scientific, Singapore.
120. Madan R. N. (1993) Chua's Circuit: A Paradigm for Chaos, World Scientific, Singapore.
121. Mancilla-Aguliar J. L. (2000) A Condition for the Stability of Switched Nonlinear Systems. IEEE Transactions on Automatic Control 45: 2077–2079.
122. Matveev A. S., Savkin A. V.(1999) Existence and Stability of Limit Cycles in Hybrid Dynamic Systems with Constant Derivatives. Part 2: Applications. In: Proc. of the 38th CDC, Japan, 4386–4391.
123. Matveev A. S., Savkin A. V. (2000) Qualitative Theory of Hybrid Dynamic Systems, Birkhauser, Boston.
124. Melander B., Bjorkman M., Gunningberg P. (2000) A New End-to-End Probing and Analysis Method for Estimating Bandwidth Bottlenecks. IEEE Global Internet Symp.
125. Melander B., Bjorkman M., Gunningberg P. (2002) Regression-Based Available Bandwidth Measurement. Int'l Symp. Perl. Eval. Comp. and Telecommun. Sys.
126. Menezes A., Van Oorschot P., Vanstone S. (1997) Handbook of Applied Cryptography. Boca Raton, FL: CRC.
127. Michel A. N., Hu B. (1998) A Comparison Theory for Stability Analysis of Discontinuous Dynamic Systems–Part I: Results Involving Stability Preserving Mappings. In: Proc. of the 37th CDC. 1635–1640.
128. Michel A. N. (1999) Recent Trends in the Stability Analysis of Hybrid Dynamic Systems. IEEE Transactions on Circuits and Systems-I: Fundamental Theory and Applications 45: 120-134.
129. Microsoft Inc. Netshow Service, Streaming Media for Business. http://www.microsoft.com/NTServer/Basics/NetShowServices.

130. Mori Y., Mori T., Kuroe Y. (1997) A Solution to the Common Lyapunov Function Problem for Continuous–Time System. In: Proc. of the 36th CDC. 3521–3522.

131. Mori Y., Mori T., Kuroe Y. (1998) On a Class of Linear Constant Systems Which Have a Common Quadratic Lyapunov Function. In: Proc. of the 37th CDC. 2808–2809.

132. Morse A. S., Mayne D. Q., Goodwin G. C. (1992) Applications of Bysteresis Switching in Parameter Adaptive Control. IEEE Transactions on Automatic Control 37: 1343–1354.

133. Morse A. S. (1996) Supervisory Control of Families of Linear Set–Point Controllers– Part 1: Exact Matching. IEEE Transactions on Automatic Control 41: 1413–1431.

134. Morse A. S. (1997) Control Using Logic–Based Switching, Springer–Verlag, London.

135. Morse A. S., Pantelides C. C., Sastry S. S., Schumacher J. M. Editors (1999) Introduction to the Special Issue on Hybrid Systems. Automatica 35: 348–449.

136. MSRA SVC software packege, http://mpeg.nist.gov/cvsweb/

137. Nandagopal T., Venkitaraman N., Sivakumar R., Bharghavan V. (2000) Delay Differentiation and Adaptation in Core Stateless Networks. Proceedings IEEE INFOCOM.

138. Narendra K. S., Balakrishnan J. (1993) Improving Transient Response of Adaptive Control Systems Using Multiple Models and Switching. In: Proc. of the 32nd CDC. 1067–1072.

139. Narendra K. S., Balakrishnan J. (1994) A Common Lyapunov Function for Stable LTI Systems with Commuting A–Matrices. IEEE Transactions on Automatic Control 39: 2469–2471.

140. Narendra K. S., Balakrishnan J. (1997) Adaptive Control Using Multiple Models. IEEE Transactions on Automatic Control 42: 171–188.

141. Ogorzale M. J., Dedieu H., (1997) Identifiability and Identification of Chaotic Systems Based on Adaptive Synchronization, IEEE Transactions CAS-I 44: 948–962.

142. Ohm J. R. (1994) Three-Dimensional Subband Coding with Motion Compensation. IEEE Transactions on Image Processing 3: 559-571.

143. Ohm J. R. (2003) Applications and Requirements for Scalable Video Coding. ISO/IEC JTC1/SC 29 WG 11/MPEG2004/W5580, Tronheim.

144. Ooba T., Funahashi Y. (1994) Two Conditions Concerning Common Quadratic Lyapunov Functions for Linear Systems. IEEE Transactions on Automatic Control 42: 719–721.

145. Ooba T. Funahashi Y. (1994) On a Common Quadratic Lyapunov Functions for Widely Distant Systems. IEEE Transactions on Automatic Control 42: 1697–1699.

146. Oppenheim A. V., Willsky A. S. (1983) Signals and Systems, Englewood Cliffs, N.J. Prentice-Hall.

147. Orchard T. M., Sullvian G. J. (1994) Overlapped Block Motion Compensation: An Estimation-Theoretic Approach. IEEE Transactions on Image Processing 3: 693–699.

148. Ott E., Grebogi C., Yorke J. A. (1990) Controlling Chaos. Physical View Letter 64: 1196–1199.

149. Parekh A. V., Gallager R. G. (1993) A Generalized Processor Sharing Approach to Flow Control in Integrated Service Networks: the Single Node Case. IEEE/ACM Transactions on Networking 1:344–357.

150. Parekh A. V., Gallager R. G. (1994) A Generalized Processor Sharing Approach to Flow Control in Integrated Service Networks: the Multiple Nodes Case. IEEE/ACM Transactions on Networking 4:137–150.
151. Parlitz U., Chua L. O., Kocarev L., Halle K. S., Shang A. (1992) Transmission of Digital Signals by Chaotic Synchronization. Int. J. Bifurc. Chaos 2: 973–977.
152. Pecora K. L., Carroll T. L. (1990) Synchronization in Chaotic Systems, Physical Review Letter 64: 821–824.
153. Peleties P., DeCarlo R. (1991) Asymptotic Stability of m–Switched Systems Using Lyapunov–Like Functions. In: Proc. American Control Conf. 1679–1684.
154. Perez G., Cerdeira H. A., (1995) Extracting Messages Masked by Chaos, Phys. Rev. Lett. 74: 1970-1973.
155. Perkins S. P., Kumar P. R. (1989) Stable, Distributed, Real–Time Scheduling of Flexible Manufacturing/Assembly/Disassembly Systems. IEEE Transactions on Automatic Control 34: 139–148.
156. Pettersson S., Lennartson B. (1996) Stability and Robustness for Hybrid Systems. In: Proc. of the 35th CDC. 1202–1207.
157. Prasad R., Dovrolis C., Murray M., Claffy K. C. (2003) Bandwidth Estimation: Metrics, Measurement Techniques, and Tools. IEEE Network 17: 27–35.
158. Radha H., Schaar M. van der, Y. Chen (2001) The MPEG-4 Fine-Grained Scalable Video Coding Method for Multimedia Streaming over IP. IEEE Tans. on Multimedia. 3: 52–68.
159. Radha H., Schaar M. van der (2001) A Hybrid Temporal-SNR Fine-Grained Scalability for Internet Video. IEEE Transactions on Circuit and Systems for Video Technology. 11. 318–331.
160. Samoilcnko A. M., Perestyuk N. A. (1995) Impulsive Differential Equations, World Scientific, Singapore.
161. Savkin A. V. (1998) Regularizability of Complex Switched Server Queueing Networks Modelled by Hybrid Dynamic Systems. System and Control Letters 35: 291–299.
162. Savkin A.V., Matveev A. S. (1998) Cyclic Linear Differential Automata: A Simple Class of Hybrid Dynamic Systems. In: Proc. of the 37th CDC. 3214–3217.
163. Savkin A. V., Skafidas E., Evans R. J. (1999) Robust Output Feedback Stabilizability via Controller Switching. Automatica 35: 69–74.
164. Schneier B. (1996) Applied Cryptography: Protocols, Algorithms, and Source Code in C (Second edition). J. Wiley, New York.
165. Schwarz H., Marpe D., Wiegand T. (2004) Scalable Extension of H.264/AVC. ISO/IEC JTC1/SC29 WG11 MPEG2004/M10569/S03. Munchen, Germany.
166. Schwarz H., Wiegand T., Lee M. H. (2004) Working Draft 1.0 of ISO/IEC 21000-13 Scalable Video Coding. ISO/IEC JTC 1 SC 29/WG 11/N6519.
167. Schweizer J., Kennedy M. P. (1995) Predictive Poincare Control: a Control Theory for Chaotic Systems. Phys. Rev. E. 52: 4865–4876.
168. Seron M. M., Hill D. J. (1995) Input-Output and Input-to-State Stabilization of Cascaded Nonlinear Systems. In: Proceedings of the 34th IEEE CDC. 4259–4264.
169. Shim H., Noh D. J., Seo J. H. (1998) Common Lyapunov Function for Exponentially Stable Nonlinear Systems. In: the 4th SIAM Conf. on Control and its Application. 323–328.
170. Short K. M. (1994) Steps Toward Unmasking Secure Communication, Int. J. Bifurc. Chaos 4: 959–977.
171. Short K. M. (1996) Unmasking a Modulated Chaotic Communications Scheme, Int. J. Bifurc. Chaos 6: 611–615.

172. Shorten R. N., Narendra K. S. (1997) A Sufficient Condition for the Existence of a Common Lyapunov Function for Two Second–Order Linear Systems. In: Proc. of the 36th CDC. 3521–3522.

173. Shorten R. N., Narendra K. S. (1998) On the Stability and Existence of Common Lyapunov Functions for Stable Linear Switching Systems. In: Proc. of the 37th CDC. 3723–3724.

174. Shreedhar M., Varghese G. (1996) Efficient Fair Queuing Using Deficit Round Robin. IEEE/ACM Transactions on Networking 4: 375–385.

175. Sigeti D. E. (1995) Exponential Decay of Power Spectra at High Frequency and Positive Lyapunov Exponents. Physical Letter D 82: 136–153.

176. Sliva C. P. (1993) Shil'nikov's Theorem - A Tutorial. IEEE Transactions on Circuits and Systems Part I: Fundamental Theory and Applications 40: 675–682.

177. Sontag E. D., Wang Y. (1995) On Characterization of Input-to-State Stability Property. Systems Control Letters 24: 351 – 359.

178. Stilwel D. J., Rugh W. J. (1999) Interpolation of Observer State Feedback Controllers for Gain Scheduling. IEEE Transactions on Automatic Control 44: 1225–1229.

179. Stiver J., Antsaklis P. J. (1992) Modelling and Analysis of Hybrid Control Systems. In: Proc. of the 31th CDC. 3748–3751.

180. Stocia I., Shenker S., Zhang H. (1998) Core–Stateless Fair Queueing: Achieving Approximately Fair Bandwidth Allocations in High Speed Network. Technical Report CMU-CS-98-136, Carnegie Mellon University.

181. Sullivan G., Wiegand T. (1998) Rate Distortion Optimization for Video Compression. IEEE Signal Processing Magazine 74-90.

182. Sutton R. J. (2002) Secure Communication: Application and Management, J. Wiley, New York.

183. Tillier C., Popescu B. P., Schaar M. V. der (2003) 3-band Temporal Lifting for Scalable Video Coding. In: IEEE Int. Conf. on Image Processing.

184. Tillier C., Popescu B. P., Schaar M. V. der (2004) Motion-Compensated Triadic Temporal Scalable Codec-A Standalone Tool in Response to The Call For Proposals on Scalable Video Coding Technology. ISO/IEC JTC1/SC29/WG11 MPEG2004/M10569/S23 (M10538). Munchen.

185. Turaga D. S., Van del Schaar Mihaela (2002), Unconstrained Motion Compensated Temporal Filtering. ISO/IEC JTC1/SC29/WG11 MPEG03/M8838.

186. Varaiya P. P. (1993) Smart Cars on Smart Roads: Problem of Control. IEEE Transactions on Automatic Control 38: 195–207.

187. Vetterli M. and Gall D. L. (1989) Perfect Reconstruction FIR Filter Banks: Some Properties and Factorizations. IEEE Transactions on Acoustics, and Signal Processing 37: 1057–1071.

188. Vidyasagar M. (1978) Nonlinear System Analysis, Englewood Cliffs: NJ: Prentice-Hall.

189. Wang X., Schulzrinne H. (1999) Comparison of Adaptive Internet Multimedia Applications. IEICE Transactions on Communications 17: 806-818.

190. Wei J., Soong B. H., Li Z. G. (2004) A New Rate-Distortion Model for Video Transmission Using Multiple Logarithmic Functions. IEEE Signal Processing Letters, 11: 64-67.

191. Wicks M. A., Peleties P., Decarlo R. A. (1998) Switched Controller Synthesis for Quadratic Stabilization of a Pair of Unstable Linear Systems. European Journal of Control 4: 140–147.

192. Wiegand T. and Sullivan G. J. (2003) Study of Final Committee Draft of Joint Video Specification (ITU-T Rec. H.264—ISO/IEC 14496-10 AVC). In: 7th JVT Meeting, JVT-G050.

193. Wiegand T., Schwarz H., Joch A., Kossentini F., Sullivan G. J. (2003) Rate-Contrained Coder Control and Comparison of Video Coding Standards. IEEE Transactions on Circuits and Systems for Video Technology 13: 688-703.
194. Witsenhausen H. S. (1966) A Class of Hybrid–State Continuous–Time Dynamic Systems. IEEE Transactions on Automatic Control 11: 161–167.
195. Wong M. K., Bonomi F. (1998) A Novel Explicit Rate Congestion Control Algorithm. In GLOBECOM'98. 2125–2132.
196. Wu C. W., Chua L. O. (1993) A Simple Way to Synchronize Chaotic Systems with Applications to Secure Communication Systems. Int. J. Bifurc. Chaos 3: 1619–1627.
197. Wu F., Li S. P, Zhang Y. Q. (2000) DCT-Prediction Based Progressive Fine Granularity Scalability. In: Proc. IEEE International Conference on Image Processing 556-559.
198. Xie W. X., Wen C. Y., Li Z. G. (2000) Impulsive Control for the Stabilization and Synchronization of Lorenz Systems. Physical Letters A 275: 67–72.
199. Xu J. Z., Xiong Z. X., Li S. P., Zhang Y. Q. (2001). Three-Dimensional Embedded Subband Coding with Optimized Truncation (3-D ESCOT). Applied and Computational Harmonic Analysis 10:290-315. I
200. Xu J. Z., Xiong R. Q., Feng B., Sullivan G., Lee M. C., Wu F., Li S. P. (2004). 3D Sub-band Video Coding Using Barbell Lifting. ISO/IEC JTC1/SC 29 WG 11 MPEG2004/M10569/S05, Munchen, Germany.
201. Xu X. H., Li Z. G., Li Y. P. (1996)Generalized Petri Nets for a Class of Hybrid Dynamic Systems. In: the 13th IFAC World Congress. 305–311.
202. Yang T., Wu C. W., Chua L. O. (1997) Cryptography Based On Chaotic Systems. IEEE Transactions on Circuits and Systems Part I: Fundamental Theory and Applications 44: 69–472.
203. Yang T., Yang L., Yang C. (1997) Impulsive Synchronization of Lorenz Systems. Physical Letters A 226: 349–354.
204. Yang T., Chua L. O. (1997) Impulsive Stabilization for Control and Synchronization of Chaotic Systems: Theory and Application to Secure communication. IEEE Transactions on Circuits and Systems Part I: Fundamental Theory and Applications 44: 976–988.
205. Yang T. (1999) Chaotic Secure Communication Systems: History and New Results. Telecommunications Review 44: 597-634.
206. Yang T. (1999) Impulsive Control. IEEE Transactions on Automatic Control 44: 1081–1083.
207. Yang X. K., Lin W. S., Lu Z. K., Ong E. P., Yao S. S. (2003) Perceptually-Adaptive Hybrid Video Encoding Based On Just-noticeable-distortion Profile. In: SPIE 2003 Conference on Video Communication and Image Processing (VCIP), Lugano, Switzerland 1448-1459.
208. Ye H., Michel A. N., Hou L. (1995) Stability Theory for Hybrid Dynamic Systems. In : Proc. of the 34th CDC. 2679–2684.
209. Ye H., Michel A. N., Hou L. (1996) Stability Analysis of Discontinuous Dynamic Systems with Applications. In: the Proc. of the 13th IFAC World Congress. 461–466.
210. Ye H., Michel A. N., Hou L. (1996) Stability Analysis of Systems With Impulse Effects. In: Proc. of the 35th CDC. 159–160.
211. Ye H., Michel A. N., Hou L. (1998) Stability Theory for Hybrid Dynamic Systems. IEEE Transactions on Automatic Control 43: 461–474.
212. Ye H., Michel A. N., Hou L. (1999) Stability Analysis of Systems with Impulse Effects. IEEE Transactions on Automatic Control 43: 1719–1723.

213. Yuan X. J., Li Z. G., Soong B. H., Wen C. Y. (2001) A RAM/DAM Switching Mechanism for ATM ABR Traffic Control. Computer Communication 24: 622–630.
214. Yuang M. C., Liang S. T., Chen Y. G. (1998) Dynamic Video Playout Smoothing Methid for Multimedia Applications. Multimedia Tools and Applications 6: 47-59.
215. Zefran M., Burdick J. W. (1998) Design of Switching Controllers for Systems with Changing Dynamics. In: Proc. of the 37th CDC, December 14–20, 2113–2118.
216. Zhang L., Deering S., Estrin D., Shenker S., Zappala D. (1993) RSVP: A New Resource ReSerVation Protocol. IEEE Networks 7: 8–18.
217. Zhang M. J., Tarn T. J. (2003) A Hybrid Switching Control Strategy for Nonlinear and Underactuated Mechanical Systems. IEEE Transactions on Automatic Control 48: 1777-1782.
218. Zhou H., Ling Y. T. (1997) Problems with the Chaotic Inverse System Encryption Approach, IEEE Transactions on Circuits and Systems Part I: Fundamental Theory and Applications 44: 268–271.

Index

Lecture Notes in Control and Information Sciences

Edited by M. Thoma and M. Morari

Further volumes of this series can be found on our homepage:
springeronline.com